Thomas A Clark

Workshop Notes and Sketches for Handicraft Classes

Being a first year's course in wood and metal working

Thomas A Clark

Workshop Notes and Sketches for Handicraft Classes
Being a first year's course in wood and metal working

ISBN/EAN: 9783337094614

Printed in Europe, USA, Canada, Australia, Japan

Cover: Foto ©Paul-Georg Meister /pixelio.de

More available books at **www.hansebooks.com**

WORKSHOP NOTES & SKETCHES

FOR

HANDICRAFT CLASSES ·

BEING

A FIRST YEAR'S COURSE

IN

WOOD & METAL WORKING

BY

THOMAS A. CLARK, M.I.MECH.E.

SUPERINTENDENT OF WORKSHOPS AND TECHNICAL DRAWING MASTER,
GEORGE HERIOT'S HOSPITAL SCHOOL, EDINBURGH.

SECOND EDITION
REVISED AND ENLARGED

EDINBURGH
JAMES THIN, PUBLISHER TO THE UNIVERSITY
1892

PREFACE.

TECHNICAL EDUCATION being of so recent introduction
into our schools, teachers have meantime at their command
only a limited selection of text-books on the subject, and
in one of its branches, handicraft, no text-book entirely
suitable for young pupils has yet been published.

That the present book, which is the result of several
years' experience in the teaching of handicraft in schools,
will altogether supply the want, it may be too much to
suppose; but so far as it goes, the course which it lays
down has been found to yield very satisfactory results.

By providing pupils with an outline of the session's
work, and a concise description of the tools employed, a
considerable economy of time is effected.

Only the tools absolutely necessary for the various
processes are here described, but should the teacher
desire to refer to additional tools or otherwise vary the
course, advantage can be taken of the blank pages which
have been inserted chiefly for extension notes.

There may be a variety of opinion about the most
suitable practice for beginners, but in the writer's
experience, that is best which gives every one the
opportunity of becoming familiar with some common
tools and fundamental processes involving their use.

Such a training leads to manual dexterity and habits
of accurate and intelligent observation.

The exercises chosen are of a practical nature, so that any one requiring them for trade purposes may have nothing to unlearn.

To prevent confusion, dimensions have not been noted on the sketches ; they should be given to suit the materials chosen for practice. Frequent reference should be made to the proportion of all parts in the various exercises, and examples of their application exhibited or enumerated.

It may only be added that combinations will readily suggest themselves to the experienced teacher, which will enable him to vary, and to make additions to the course of work here sketched.

PREFACE TO THE SECOND EDITION.

THIS Edition of "Workshop Notes and Sketches" has been carefully revised and considerably enlarged. In response to numerous requests made to him, the Author has added an entirely new part on " Metal Work," which it is hoped may meet with an equal measure of approval to that bestowed on his Wood Work notes.

Under the head of Metal Work, the subjects of turning lathes, planing, shaping and drilling machines, blacksmithing, moulding, and fitting are taken up and treated from the standpoint of a practical teacher, whose daily duty it is to instruct in the various branches of handicraft dealt with in this little volume.

October 1892.

CONTENTS.

PART I.
WOOD WORKING TOOLS.

PART II.

WOOD WORKING.

PART III.

METAL WORKING.

SKETCHES.

PART I.

GENERAL DESCRIPTION OF TOOLS.

Bench Planes.—An ordinary *set* comprises, in general the three—(a) *jack plane*, (b) *halflong, truing or trying plane*, and (c) *hand* or *smoothing plane*. The jack plane is used for removing the rough surface after the saw, securing approximately correct surfaces, and for rough work generally. The halflong is employed for straightening, flattening or truing surfaces after the jack plane. The hand plane is used for smoothing or finishing work after the other planes, as also for numerous purposes for which the jack or halflong would be unsuitable. The *stocks* of bench planes are made of *beech*. The *irons* or cutters are of *wrought iron* faced with *cast steel* to form the cutting edge, which makes grinding more easily performed than if they were wholly made of cast steel. The irons are fixed in the planes at the angle of 45° to the face and kept in position by beech wedges Plane irons are sharpened on the grindstone at the angle of about 25°, on one side only—the other being left quite flat; and afterwards on the oilstone at about the angle of 35°, thus leaving about 10° as an *angle of clearance*. Each iron is provided with a steel *cover* (kept in position by a screw) for the purpose of strengthening the cutting edge, and at the same time acting on the shaving as it is being removed, raising it up at a more obtuse angle and enabling the cutting edge to get closer to the point where the shaving is being severed from the wood. This secures a better cutting action, and removes to a great degree the splitting tendency, otherwise unavoidable. The screws are all *right hand*, and are tightened on being turned by the thin *pane* of a pin hammer, in the same direction as that of the forward motion of the hands of a clock or watch. The cover of a jack plane iron should be kept about $\frac{1}{16}''$ from the cutting edge, whilst those of halflong and hand planes should be placed as close to the edge as practicable.

To keep planes in good working order, the stocks, when new, ought to be well soaked with *raw linseed oil*, and receive an occasional coat afterwards. This will cause them to work more easily, last longer, look better, and be more pleasant to handle.

Saws.—The saws required in general work are the *rending* or *ripping-saw*, *cross-cut* or *hand-saw*, *panel-saw*, *tenon-saw*, *dovetail saw*, and *compass-saw*. They may be classified into three divisions, *(a)* for cutting along the grain of the wood, *(b)* for cutting across the grain and *(c)* for cutting both ways. The rending-saw is the only one belonging to the first class, the cross-cut and panel-saws belong to the second, and the remaining three to the third. The chief differences exist in form and size of teeth and blade.

RENDING-SAW.—*The cutting faces of rending-saw teeth are almost at right angles to a line touching their points.* The teeth are nearly of the form of right angled triangles, but for hard woods this *hook* or angle is too great, and the teeth are sloped backwards so as to make the angle of contact with the wood more obtuse.

CROSS-CUT OR HAND-SAW.—This saw has its teeth more in the form of equilateral triangles, and thus resembles a rending saw used for hard wood, the angle being nearly equal at both sides. The teeth of this saw are somewhat smaller than those of the rending saw, as the saw itself is also smaller than the other.

PANEL-SAW.—The panel-saw is very similar to the cross-cut, only its teeth are smaller, and its blade smaller and thinner.

TENON-SAW.—This saw has a thinner blade and finer teeth than any of the foregoing, and is used for finer work. To strengthen the thin blade, a thick *back* is supplied, giving to it the name of the *back-saw*.

. DOVETAIL-SAW.—This is like the tenon-saw in all respects, only smaller and finer every way, and is used for finer and lighter work.

COMPASS-SAW.—The compass-saw has a narrow, tapered blade, and is used for curved work, as in following the lines traced by a compass.

Saw blades are made of cast steel plate, while the backs of tenon and dovetail-saws are made of iron, steel, or brass. The handles are made of beech, fixed to the blades by brass screws. Good saws have their blades ground thinner towards the back. This makes them more easily worked, even with a smaller amount of *set*.

SET.—Set is the bending of adjacent teeth to opposite sides, to make a wider *saw-kerf* and thus reduce the friction on the sides of the saw. The cleaner the saw.cuts and the drier the wood, the less necessity for set ; but if the wood is wet, or the saw *dull*, then the value of set will be more readily perceived.

SHARPENING.—Saws are sharpened by files of triangular section, the blade being secured in a suitable vice. Alternate teeth are sharpened from the same side, one complete side being sharpened and then the saw reversed for the sharpening of the

A

opposite side.   The file must be held at different suitable angles
to give the proper form to the teeth.

The Rule.—For workshop purposes the rule varies in
length from 6 in. to 4 ft., the longer ones being made to fold up
into convenient lengths.   The 1, 2, and 3 foot rules are most
largely used—the latter two being folded up into 6 in. and 9 in.
lengths respectively.   Each foot is divided into inches and any
suitable fractional part of an inch, as $\frac{1}{2}$ in., $\frac{1}{4}$ in., $\frac{1}{8}$ in., $\frac{1}{16}$ in., &c.
Rules are mostly made of boxwood and mounted with brass ; but
small ones, to be used also as straightedges, are often of steel or
brass.

The Square.—This tool is made up of two parts—the *stock*
and the *blade*, fixed together at the angle of 90°.   The former part
is made of rosewood, covered on the edge by a brass plate to
prevent wear, the latter is a thin steel plate, and both are fixed
together by means of rivets to prevent shifting.

Pin-Hammer.—The form of this tool for bench work is
varied, but a convenient shape is one with a thin *cross-pane*,
suitable for moving the plane iron screws.   A useful weight is
about 10 oz.   The *head* is made of cast steel and the handle of
ash.   The handle is slightly reduced in section towards the head
so as to afford it more *spring*.

Marking Gauge.—One common form of construction of
this tool is a sliding block of beech, fitted tightly on a parallel bar
of the same material and provided with a screw of boxwood, let
into the side of the former, so that it can be readily *pinched* on to
the latter, to fix them in any desired position.   At one end of
the bar, a sharp steel pin is passed through for the purpose of
marking the wood to be gauged.   This tool is employed for
gauging wood to any desired size, such as, for example, width
or thickness of a rectangular block, preparatory to planing down
to the required dimensions.

Mortise Gauge.—The mortise gauge is very much like the marking gauge, the chief difference lying in its being supplied with two marking pins instead of one. The block and bar are made of rosewood, and the screw of steel. One of the pins can be moved along the bar by means of a screw and slide, so that the marking pins may be set at any desired distance apart. This gauge is used for making two marks at one time, necessary as in the case of the mortise and tenon joint.

Paring Chisels.—These are divided into two classes, namely, *firmer chisels* and *bevel-edged chisels*. Firmer chisels are of rectangular section, while the others, as their name indicates, have their edges bevelled away at an angle, making them more suitable for some kinds of work, as also lighter to use. They are made entirely of cast steel, the cutting edges being at one end and a tapered tang at the other. The cutting edge is formed by grinding at the angle of 20° and setting on the oilstone at an angle of about 30°. This is done on one side only, the tool being left quite flat on the other, just as in the plane irons, and unlike the axe, which is sharpened equally from both sides. A hardwood handle is driven on to the tang. Chisels vary in width from $\frac{1}{16}$ in. to 2 in. and upwards, and are employed to form flat surfaces, on many of which the planes could not be applied.

Socket Chisels.—For heavier work this chisel is more suitable than the firmer chisel, and instead of the handle being driven on to a tang, it is fitted into a socket. They are also known as *mortise chisels* because much of the work done by them is of the nature of mortising. The most of their work being heavy, a hammer or mallet is necessary to drive them into the wood, the mallet being the better of the two for this purpose.

Gouges.—Gouges are very much like firmer chisels, only they have a curved form of section. They are of two kinds, namely, *firmer gouges*, which are ground on the outside of the curvature, and *incanneled gouges*, ground on the inside. This is to allow of a greater variety of work being done by them. They are to be had from ⅛ in. upwards. The general remarks on firmer chisels apply here, except that in sharpening the concave surface, a *slip* of oilstone with a suitably curved edge has to be used, the oilstone itself being employed for the outside.

Oilstone.—This piece of bench-furnishing has a variety of names, such as *hone, set-stone, whet-stone, sharpening-stone*, &c. It is used for sharpening or setting edge tools after the grindstone has performed its part. The oilstone ought to be mounted in a small case of hard wood, having its upper surface covered when not in use, to protect from dust. Some good oil, such as sperm, should be applied on the surface operated upon, which surface ought to be kept as clean and flat as possible.

Slip Stone.—This is simply a small piece of oilstone with rounded edges of different thickness and curvature, and is used for sharpening gouges or edge tools of curved or irregular form, on which the common oilstone cannot be conveniently brought to act. It may be fixed in a case with the edge up, or used in the hand without any case.

Oil Can.—The most suitable form for bench use is made of conical shape, either of brass or tinplate. The bottom is flexible, and a nozzle having a small hole is screwed into the top. The discharge of oil is effected by inverting the *can* and lightly pressing the bottom.

Paring Board.—To prevent the benches from being cut with chisels, or other edge tools, and likewise to prevent the tools themselves from being blunted by coming in contact with any sand, nail, or other hard substance which might be on the bench-top, it is advisable to be provided with a small piece of service wood to work upon, which should be used at all times when any process entails such operations as paring, &c. Its top surface should be kept clean and flat, and if damaged in such a way as to mark the work, should be refaced.

Bench-Hook.—Although much of the sawing generally done against the bench-hook can be performed against the *bench stop*, yet if the wood has to be cut *through*, it is the safer way to use the hook, as there is less danger of cutting the surface of the bench. Any cutting of the bench is always objectionable and more especially near the stop itself.

Sliding Bevel.—Although different from the square, still the sliding bevel is used for exactly similar work. Its corresponding parts are made of the same materials as those of the square, only the stock and blade are fixed together by means of a screw instead of rivets, which allows them to be set at any desired angle to one another. The blade is provided with a slot, so that it may be shifted on end to suit special circumstances.

Marker.—This tool is also known by such names as the *drawpoint* and *scriber*. It is formed of a tapered piece of steel wire about 6 in. long, and ¼ in. diameter at one end (on which may be fixed a handle) but coming to a sharp point at the other. It is employed for marking or lining the more exact pieces of work, such as in dovetail jointing, &c., when it is used to trace the exact size and form of the dovetail pins upon the piece in which the holes are to be made, or to make other exact transfers when the pencil would not be suitable.

Mallet.—When mortising or heavy paring has to be done, it is not advisable to use a hammer, as the tool handles are too readily damaged by this means. For such purposes the mallet should be taken. Like hammers, these tools are made in a variety of shapes, guided by the kind of work and the taste of the workman. They may be made of various hard woods, the chief item of importance being to select a piece of wood of tough nature and not ready to split.

Wing Compass.—The chief difference between this one and the simple compass is that the former is provided with a radial wing fixed into one leg and passing through a small slot in the other, where it can be pinched by a thumb screw to set the compass at any required radius. Besides being used for describing circles, or parts of circles, on the work under treatment, one common use this tool is put to is that of laying off equal distances either in a straight line or along the circumference of a circle. By this means a much more exact result is arrived at, than by the employment of the rule and pencil.

Brad Awls.—Brad awls are used for making holes suitable for small nails and screws. They are made of steel and vary in diameter from the size of a small needle up to $\frac{1}{4}$ in. They are sharpened equally at opposite sides and brought to a sharp flat point. On the other end there is a tang on which is driven a handle. The cutting edge should be inserted into the wood across the grain to prevent splitting. The tool ought to be held as nearly as possible in the direction in which the hole is begun, as any attempt to alter it after the hole has been nearly bored might result in the breaking of the awl. Equal care should be taken in extracting it from the wood, for the same reason.

Screw Driver.—Another name for this tool is the *turn-screw*. It is used for turning or driving screw nails, as its names suggest, and is made of suitable sizes to fit different screws. The bar which acts on the screw is made of steel and thinned on the point to fit the groove in the head of the screw. Like the chisel, the other end is formed into a tang, upon which is fixed a handle somewhat broader and flatter than the chisel handle.

Pincers.—The pincers are used to pull out service nails, or those which have gone wrong in driving. Although quite suitable for bench work, they are discarded by joiners in roof work, floor work, and general lining, in favour of the *claw-hammer*, which can perform the double function of both driving and drawing nails.

Spokeshave.—For finishing many small curves the planes are quite unsuited, and have to give place to the spokeshave, which is constructed for the purpose. Boxwood and beech are the usual materials of which its stock is made, but metal is also used. The iron has two tangs at right angles to the cutting edge, by means of which it is held firmly in the stock, the position of the iron regulating the thickness of the shaving, as in the bench planes. Both hands are needed to work the spokeshave, and the guiding surface being so small, makes dexterity in its use rather difficult to acquire.

PART II.

NOTES AND SKETCHES OF EXERCISES IN WOODWORK.

Sawing or Breaking out Wood.—Before commencing to cut out material for any piece of work, it is necessary to know exactly the sizes of the several pieces required, the best method of construction, and the class of wood to be used. Unless these points are attended to, mistakes will be the outcome, leading to loss of time and material, besides the danger of getting into bad habits. It is always advisable to see the way clear from the beginning to the end, and this can only be done by having correct drawings of the articles to be constructed, from which a thorough mental grasp of the whole should be obtained before starting work. When a list of the sizes and sorts of wood required, with allowance for sawing and planing, has been made up and the wood selected, the rule, pencil, square, straightedge, and saws should be at hand for lining on and cutting out the pieces required. If those pieces are parallel, and one edge of the board from which they are to be taken is quite straight, then a very simple way of drawing the lines parallel to the edge, is by using the rule and pencil like a marking gauge. The rule, held in the left hand, is placed square across the top surface of the board as far as the desired width of the piece to be cut out, and kept in position by the thumb nail, which performs the functions of the sliding block of the marking gauge. The pencil, as marker, must be pressed against the end of the rule, and be in contact with the surface of the wood. If both rule and pencil are now simultaneously and carefully moved towards the body, a line, parallel to the edge, will be drawn. By adopting this practice, a considerable amount of time is saved. The forefinger nail of the left hand should not be used in place of the thumb nail, for fear of accident. In many instances the wood may not be straight on the edge, and the pieces required may be of irregular shapes, and in these the rule and straightedge should be resorted to. The best arrangement of the pieces to be cut from a board, is that which allows them to be *broken out* most easily and with the smallest possible waste. Should knots or small cracks be objectionable, then both sides of the wood ought to be well examined before lining, to avoid disappointment, and waste of time and material when the work has reached a more advanced stage. This is a safe precaution at all times.

In *sawing out*, the greatest care should be observed to cut exactly by the lines and quite square through, but never on any account to cut further than is absolutely necessary. As considerable difficulty is generally experienced by beginners in starting the cut, it will be found a great assistance if the weight of the saw is nearly all borne by the right hand, while the points of the teeth are brought lightly in contact with the wood, the blade being guided into position by the thumb of the left hand. Should the saw tend to leave the line, this can be counteracted by a slight twist on the handle in the opposite direction, by the right hand. This tendency is often due to a blunt side of the saw, or less *set* on one side than the other, but more generally to a constant pressure by the hand in one direction, which should be early avoided.

Planing.—In using the planes, so as to obtain good results, they ought to be kept quite sharp; the covers must be screwed firmly in their proper places on the irons; the shavings should be kept clear out of the plane mouth to prevent choking; and the planes held steadily and straight along the wood.

Planing the Face.—The best surface should be selected for the face. If a piece of wood has been standing for some time on the ground, the end may be covered with sand, and to avoid blunting the planes, it may either be cut off with the saw, or a small corner, called a *chamfer*, taken off the surface to be planed This may be done by the paring chisel or jack plane. A *shaving* should then be taken off all over, and the face made approximately correct with the jack plane before using the halflong upon it. The properties of a true face are threefold, 1st, flat across, tested by the edge of the square blade applied crossways; 2nd, straight lengthways, tested (if short) by the edge of the square blade placed along the wood, but if long, by a straightedge of suitable length, or more commonly by the eye; and 3rd, out of twist or *winding*, tested by diagonal application of the edge of square blade both ways, by the eye, or by two *winding-sticks*, which are simply parallel pieces of wood laid across the piece to be tested, one at each end, with the edge up, when any twist or winding may be detected by looking across their upper edges. After the face has been properly finished with the halflong, some distinctive mark should be put upon it, as on A, Fig. 12.

The Edge.—If the wood is thin and broad, it will be most easily held in position by fixing in the vice, with the edge up, which will allow both hands to be free for manipulating the planes. The properties of a true edge are its being, 1st, at right angles to the face, and 2nd, straight lengthways. The

WORKSHOP NOTES AND SKETCHES. 13

former is tested by firm application of the edge of square stock
upon the *face*, with the edge of the blade passing across the *edge*
of the wood, accuracy being obtained when the wood exactly
fills the square. The latter is tested by the same means as the
corresponding property of the *face*. The *edge* when finished with
the halflong should be marked in the same way as the *face*. None
of the other sides, however, should be marked in this manner.

The Ends.—The ends can be most easily planed when the
wood is fixed into the vice, end up, and only a little above the
surface of the bench. As in this process there is a danger of
splitting pieces off the edge, it is advisable to use as little *iron* as
practicable and to plane *away* from the finished edge. To prevent
splitting, a small chamfer can be taken off the far corner with the
paring chisel. If sand should happen to be on the end, then a
short piece should be cut off with the saw before planing. The
properties of a true end are—square to both face and edge, and
tested by the square in a similar manner to the first property
of the edge. Both ends may be planed if the exact length is
known.

Gauging to Width and Thickness.—Gauging is
somewhat difficult for the beginner. The chief error is the
tendency to put too much weight on the marking point and too
little pressure of the sliding block on the face or edge. Such
mistakes result in a crooked and irregular line, not at all parallel
to the surface gauged from. A very light mark should at first be
made, seeing that the block is pressed firmly and flatly on the
surface gauged from, while the gauge should be slightly tilted to
prevent the marker getting too deep into the wood. When a
correct line has been formed, the depth can be readily increased
so as to make it distinctly visible. In planing the rough part with
the jack plane down to the gauge lines, occasional examinations
should be made, in order to insure a regular approach to the

lines, without going beyond at any point. The surfaces should be finished quite flat with the halflong to the centre of the gauge-lines, and the square blade may be applied occasionally to test the truth of each.

Modifications of the foregoing instructions will have to be made in cases when the conditions may be different, as, for instance, when the edges or ends do not require to be at right angles to the face. In such cases the sliding bevel, set at the required angle, would be used in a similar way to the square.

Jointing.—In some exercises no jointing of the several pieces may be required, as they may be fixed together by glue, nails, or screws. In construction of the great majority of articles made of wood, joints of various sorts are an absolute necessity. In order that sufficient practice may be got in planing, gauging, exact measurement, and cutting to lines preparatory to jointing, it will be advisable to procure a piece of wood about 1 in. square, and from 12 to 15 in. long. It must be planed correctly on face and edge, and gauged to the greatest size it will stand, either $\frac{7}{8}$ in., $\frac{3}{4}$ in., or under. When finished, a line must be accurately drawn on each side, at right angles to face or edge, by the square and pencil (observing the advice given at the centre of page 16), and as far from the end as the width of the wood. This part must now be carefully cut off with the dovetail saw, so that each line may be split, and the end left quite square. If not successfully performed at first, the process should be repeated till the cube is exact, after which the same thing should be done by the help of the angular groove, cut on each side, as described on page 17. If unsatisfactory, this, also, should be repeated until the work is without fault.

Halving.—This class of joint is used for a great many different purposes, and the joint itself assumes quite a number of forms, depending upon the application. Halving is much applied

in housebuilding, as also in many pieces of general framing used
in mechanical engineering, &c. Applications of this joint are
very common, as a casual observation will show. Figs. 1 and 2
are selected when the ends cannot be allowed to pass beyond the
flush of the sides, Figs. 3, 4, and 5 when one end may pass, but
the other be flush or even with the side, and Fig. 6 when both
ends may project. Figs. 2 and 4 are called dovetail halving and
Fig. 5 bevelled halving. These joints may be used for uniting
pieces either at right angles or obliquely as Figs. 7, 8, and 9.
Figs. 10 and 11 are applications of the same joint, but many
additional ones are in common use.

In making halving joints, or in fact any class of joint, one
of the chief points of the work is to have the several parts
properly proportioned, and the lines carefully drawn on. As the
same general instructions are suitable for all these joints, a
description of one, say Fig. 4, will suffice for the whole. Fig. 4
is known as dovetail halving, because of the shape of one part of
the joint, after the side pieces have been taken off. The first
thing to be done, after the wood is planed to required sizes, will
be to arrange the pieces for jointing, so as to have the edges to
the most suitable sides, and the faces to the top or same side. In
this instance, shown in detail at Fig. 12, the edge of A may be
kept inside, while that of B may be put to right or left. The
reasons for placing or arranging pieces in a certain way before
jointing, are, that certain surfaces may be to certain sides when
the work is completed, and that every advantage may be taken
of the true surfaces, to secure good and close joints. The faces
are kept to the same side, so that if any gauging has to be done,
the surfaces will be flush when the pieces are put together, as
half is taken out of each; and likewise to have all the best
surfaces to one side. In some joints it is best to make one part
first, and then make a transfer by the use of a sharp pencil, a

marker, or chisel edge. This, however, is only advisable when it
would take longer time by measurement, or be likely to prove less
correct when done.   As one of the chief aims of this sort of
education is to obtain accuracy of observation, all methods likely
to beget the opposite should be avoided.   To this end, careful
measurement should be insisted on at a very early stage, and no
guess work encouraged, or allowed to pass.   It is one of the worst
possible styles of doing work, to allow so much for fitting, and
then gradually take it off piece by piece until the proper size is
arrived at.   This must be done in some instances, but should by
no means be recognised as the right way, if it is practicable to
measure the sizes exactly.

The first line to be drawn is c d across the under side of b
(Fig. 12), square to its edge and about $\frac{1}{16}$ in. further from the end
than the width across A.   Lines c e and d f are to be drawn
square across the edges from the ends of this line, and care should
be taken *in every squaring operation to have the brass edge of the
square stock firmly placed against either face or edge.*   The last
line drawn square across the face, if correct, should exactly join
the ends, e and f, of the two lines on the edges.   The lines for
the sides of the dovetail may then be put on, allowing sufficient
taper to give strength, while keeping the neck of suitable
width.   Equal distances e g and f h are to be measured along
the line e f and the points g and h joined to the points i and j at
the extreme corners.   Although not absolutely necessary, the
same may be done on the opposite side, in order to have an
additional safeguard and guide.   This may be of no service, as
the under piece of b, on which the extra lines would be drawn,
may be cut off before the other process is performed.   If found
necessary these lines can be put on after the slab has been
removed, but as the square may be readily applied from the face,
these extra lines will be found of less importance.   Corresponding

lines G′ I′ and H′ J′ can now be put upon the face of A, and the ends of the two lines squared half down the edges. The marking gauge should then be set to half the thickness of the wood, and gauge lines *drawn off the faces*, along both edges of each piece and on the end of B. Both pieces are now ready for the making of the joint. As it is somewhat difficult to begin the saw-cuts accurately by the lines, and seeing that the saw does not leave a very smooth edge, it is better in the first place to form a small shoulder or angular groove, with the paring chisel, to insure an exact beginning. To do this, place the square close by the line, holding tightly with the left hand, while the chisel is held in the right in an upright position, to form a square edge, and with its flat side against the square blade, but tilted so as to make the corner cut easily. A light line is first drawn across, and if correct, thereafter made of suitable depth, say $\frac{1}{18}$ in. The chisel is then sloped over to the angle of about 45°, and inserted so as to reach down to the bottom of the vertical cut already made. If this groove be well formed, the tenon saw may be entered without difficulty. The piece of wood B—inverted—is shown with these grooves formed at B′, Fig 12. The same may be done across both edges of B, and also across by the oblique lines on the face of A. It will be found advantageous to resort to this process wherever a shoulder is to be made, or in fact, when any surfaces of this kind are to be fitted together, and on which the planes cannot be applied. The nature of each joint must be studied, for it may be easily spoilt by taking out the sloping surface with the chisel at the wrong side of the lines, and instead of a close joint the reverse will be the result. The tenon or dovetail saw would now be brought into requisition, and the several parts cut, while the slab, below the gauge lines on B, would be cut off, or in the case of small pieces with straight grain, they might be split with the paring chisel. Sawing will be found to be generally the better and quicker method. The bench-hook and vice will each be found suitable for holding the work during certain cuts—the former when the wood is horizontal and the latter when vertical. As the saw leaves a good enough surface for the inside of the joint, it should be the endeavour of the workman to saw so exactly as to require little or no paring down to the lines, and thus much time will be saved. The side pieces, to form the dovetail, would be taken off in a similar manner. The piece above the gauge lines on A can be split out with the chisel and then pared to the lines. Before driving the pieces together, a slight chamfer should be taken off the under corners of the dovetail, to make it enter more easily. This joint may be fixed together by glue, pins, nails, or screws. As already remarked, those joints are so much alike in point of construction that it is quite unnecessary to give a detailed description of each—all that has to be attended to, is to make sure of the sort of joint best suited for the work, and to have the parts properly proportioned and well fitted.

Notching.—This is closely connected with the last class of joint, and in fact notching and halving, in some instances, are almost synonymous terms, save that surfaces of halving joints are made flush with one another, whereas in notching, the part taken out to form the joint is often only a very small proportion of the depth of the piece of wood. Such joints are largely used in connection with flooring-joists, roofing, and framing of many sorts. Figs. 13, 14, and 15 show different methods, as used for joists and wall plates, purlins and rafters, &c. Fig. 16 is called dovetail notching, and Fig. 17 Tredgold's notch. As the manipulation of the tools, both while *drawing in* and making these joints, is identical with the corresponding process in halving, further detail is uncalled for.

Cogging.—This joint is similar to notching, only instead of the notch being carried right through, as in Fig. 14 a part or *cog* is left near the centre, as shown at Fig. 18, or to one side as in Fig. 19. The only difference in making this joint is that the sides have to be mostly cut down with the chisel (Fig. 20), as the saw can only be applied to do a small part of the work in this instance.

Rebate Joint.—The rebate joint is also known as the rabbit or check joint. Rebate joints are employed over a wide range of constructive processes. Many of them are applied to join thin pieces at an angle, as at the corner of a box. The corner may be a right angle, but this joint is also suitable for acute and obtuse angles. The pieces to be jointed should be arranged as directed in halving. The end out of which the rebate is to be taken must be lined square across the inside, at a distance from the end, as E F, Fig. 21, about $\frac{1}{16}$ in. more than the thickness of the piece A, to be jointed into it. Lines should then be drawn down the edges from each end of the cross line E F, at right angles to the face, and a gauge line run across the end and along the edges, parallel to the face and about half-way through, but more or less may be taken out, just as the use the joint may be put to requires. To enable an exact beginning to be made by the saw, as well as to give an accurately close joint at the shoulder, a small angular groove, c, Fig. 21, should be cut out by the paring chisel, with the square used as a guide, and as described in *halving*. The tenon saw would then be used to cut down to the gauge lines both ways, or if the wood be straight grained, the part to be taken out may be split off with a paring chisel. When correctly finished, the joint should be quite close at all points and at the same time square. Fig. 22 shows Fig. 21 applied as the joint at the corner of a box. Fig. 25 is used for the same purpose as Fig. 22, and possesses certain advantages over it. Both are used when the joint is away from

B

the end, as Figs. 23 and 24, Figs. 26, 27, and 28 show oblique
applications of the same joints. Such joints in boxes used for
holding liquids are generally coated with white lead before being
put together. The dovetail, as shown at Fig. 29, may be applied
here with good results instead of the process sketched in Fig. 23.
A very small amount of taper on the length of the dovetail may
be allowed to make sure of a tight fit. Such a joint may be used
for binding others together, and will do so up to a certain limit
without any fastening.

Mitre Joint.—The mitre joint is one formed by the meeting
of two pieces at a corner, on a line bisecting the right angle.
The same class of joint may be used on angles greater or less
than 90°, as in Figs. 31 and 32. The mitre joint is extensively used
in door and window making, joining of architraves, picture frames,
and all sorts of mouldings. It is also much used in cabinet and
pattern making. *Keying* is a process sometimes employed along
with this joint to strengthen it. A saw cut is let in as far as the
dotted line on Fig. 30, into which is glued a thin slip of wood.
If it is a deep joint, as the corner of a box, a few keys may be
put in, as shown in Fig. 33, while in a small picture frame,
only one would be used. A mitre square, *i.e.*, one set at the
angle of 45°, is used for drawing on the lines for this joint.
When much work of this sort has to be done, as is the case in
picture frame making, &c., a *mitre-box* is used. These assume
many different forms, some for the use of the saw only, and
others with the addition of a knife to finish the joint, or adjuncts
for the application of the plane after sawing.

Mortise and Tenon Joint.—The simplest form of this
joint is when two pieces A and B are joined together at right
angles, as in Fig. 34. The joint is formed by cutting or *mortising*
out the hole, C, Fig. 35, called the *mortise*, and on the other piece
making a part D, the *tenon*, of corresponding size, to fit tightly into

the mortise. The thickness of the tenon is generally about ⅓ the thickness of the wood, as in Fig. 36, which shows the sort of double tenon used for the bottom rail of a framed door. These joints assume a great diversity of form, because of the variety of timbers to be joined and the way they may be required to meet. Figs. 34-41 show a few joined at right angles. Fig. 42 is a modification so as to show a mitre joint either at one or both sides. Figs. 43 and 44 show the *trimmer* joint, with the *tusk-tenon*, A, used in uniting joists and girders or beams together, as in double floors or about a fire place, stair-case, or such like openings in floors. Figs. 45 and 46 show oblique application of the mortise and tenon joint as used between principal rafter and tie beam in a roof truss. The construction of the simple mortise and tenon joint is as follows :—When the position of the mortise has been fixed, and lines drawn, as well as shoulder lines for the tenon, the gauge lines have to be put on for the thickness of the tenon and width of the mortise. This is done by the mortise gauge. All the lines should be gauged from the face or edge, as in the other cases, and on both sides of the wood, if the mortise should have to be cut through. The length of the tenon can only be determined after the most suitable has been selected. If the tenon is to go through, then its length must be a little greater than the depth of the mortise, to allow a chamfer to be taken off all round the point, in order to prevent splitting at the under side of the mortise as it is driven through. This extra piece can be cut off and planed flush when the work is fixed together. A socket chisel should now be selected of suitable width, and, by help of a mallet, the part inside the mortise lines cut out. This is done by starting a little from one end with the first cut and gradually working towards the other. When the end is reached the wood must be turned over, and the same done on the opposite side, after which the loose pieces can be driven out and the mortise finished up to the lines. In making the tenon, a shoulder should be formed with the chisel, as in the halving and rebate joints, before sawing down to the gauge lines. The tenon saw would be the most suitable to form the tenon, as the name implies, and the more accurately this is done the less will be left for the chisel afterwards. For heavy work the rending saw is taken for cutting down the *cheeks*. Those joints are fixed in many instances, as indicated in Fig. 49, by wedges, A A, the mortise having been made wider at the outside to admit of this. When the tenon does not go through, the joint may either be pinned as at B, or *fox-wedged* as in Fig. 50. Mortise and tenon joints are widely employed for most articles requiring to be strongly and neatly framed together. Doors, windows, roof-trusses, many pieces of household furniture, and the majority of framed structures give examples of their use.

Bridle Joints.—Two examples of this class of fixing for timbers, used instead of the mortise and tenon joint, are shown at Figs. 47, 48. A *bridle*, or narrow bar, is left in the centre of one piece, as A, Fig. 48, about ⅕ the thickness of the wood, and the remaining parts formed similar to the oblique mortise and tenon joint, only that the space B will have to be cut out instead of being left in the form of the tenon, as already shown.

Scarfing.—When a beam has to be lengthened without enlarging the joint to any great extent, the pieces are scarfed together. Figs. 51-56 show a few different kinds of scarf joints as used for wall-plates, beams, and any long timbers. Many of these joints are strengthened by straps or plates of wood or iron, called fish plates, with through bolts to bind all together. Little need be said about the construction of the joints, as the annexed sketches should be sufficient guide. One of the main things to be attended to in making this, as in fact any joint, is to see that every part is an exact fit, and so ensure the strain being equally distributed.

Board Jointing.—When a surface is to be formed of wood greater in width than can be conveniently got, several pieces may have to be joined edge to edge, as for instance in the case of flooring. Figs. 57-61 show a few common examples. Fig. 57 shows plain or butt jointing; Fig. 58, the same with the addition of *dowel-pins* put into both edges at short intervals; Fig. 59 is a *rebated* joint; Fig. 60, A, shows the *tongue and groove* joint, while B is *rebated, tongued*, and *grooved*; Fig. 61 is a *groove* and *feather* joint, in which the groove is taken out of both the edges, and a feather either of wood or iron inserted. In flooring-boards, the tongues are kept below the centre, to allow for the wearing of the upper surface. Most of these joints are made by machinery, although plough planes, &c., are used occasionally. In laying large surfaces of wood it is the common practice to make the pieces narrow, as in flooring, so as to prevent splitting by contraction, and to show but a narrow opening at the joint should contraction take place. To remove the bad appearance of any open joint in linings, one edge is often *beaded*, as the tongue edge of Fig. 62, A, while another common practice is that of removing a small chamfer from each corner, as at B.

Dovetail Joints.—Dovetail jointing is largely used for the same purpose as the rebate joint. It is a much stronger and more durable method of union, but rather more difficult to make. A, B, C, Fig. 63, show different views of this joint in its simple form. Fig. 64 will give a clear idea of what the joint is like. Fig. 65 may indicate how the intervening parts between the pins P P, &c., are removed. A shows one drawn on ready for cutting out. The tenon or dovetail saw is taken to cut down the vertical sides of the pins to the lines, after which the pieces between the pins may be removed. This is done by cutting in with the chisel

from both sides, a little away from the line, as at B, or better
still, by means of the bow-saw (if it can be got), which has a
narrow, thin blade, and if passed down the vertical saw-cut, it
may easily be turned round and made to cut along by the line as
at C. The holes in the other piece, A, Fig. 64, would be cut out
either way, except that the outside ones should be done with the
dovetail saw.   In lining off for making this joint, lines should
be drawn round the ends, a little farther away than the thickness
of the piece to be joined thereto, allowing about $\frac{1}{16}$ in. over for
the end in which the holes are to be made, and $\frac{1}{8}$ in. more on the
other for the pins.   The pins would now be marked off on the
inside, care being taken, as with the rebate joint, that all the *faces*
are outside and the *edges* to one side.   After the widths are set off
on the inside, say half the thickness of the wood, and twice its
thickness between their centres, the square would be used to
draw lines at right angles from the ends to the cross lines, or
parallel to the edge.   From the ends of these, lines should next
be drawn across the end of the wood with the sliding bevel,
giving proper taper to the pins, and taking care not to bring them
to a sharp edge, while at the same time giving sufficient taper for
strength.   Lines would then be drawn from the ends of these,
down the face, to the cross line and parallel to the edge, or at
right angles to the end.   This would finish the drawing on
of the pins.   When all the pins are marked off, the wood
should be fixed upright in the vice, and the tenon or dovetail
saw used to cut down the side of the lines, this being done
as exactly as possible.   The large intervening portions should
now be cut out, which may be accomplished, as already explained,
by means of the bow-saw, or paring chisel and mallet.   A
very sharp chisel should be used to clean out to the lines, as at
D, Fig. 65, and the surface at the bottom should be slightly
hollow, so as to ensure close bearing on the outsides, as at C,

Fig. 66, which is a section on A B. All other surfaces should be as flat as possible to ensure a correct fit, and if the saw has been carefully used, very little should be left for the paring chisel to do.

When the pins are finished, their exact size and forms must be transferred to the end where the holes are required to be made. This is accomplished by placing the piece, just described, with its pin points resting on the inside of the other piece of wood, the inner surface of the pins touching the line formerly drawn on the ends of the said piece, and the edges flush. Both should be held firmly in this position, and lines traced exactly round the pins by the *marker*. The lines should now be examined and produced to the end, any error being corrected; then they should be carried square over the end of the wood, and those drawn on the inside should be reproduced on the *face* by the aid of the sliding bevel. This is to ensure the holes being the same size and shape at both sides, but it will become unnecessary after proficiency in using the tools is arrived at. Everything being now ready for cutting out the holes, the wood is to be fixed in the vice, end up, as in cutting the pins, and with dovetail or tenon saw, cut down exactly *inside* each set of lines. For cutting out the useless portion, the same remarks apply as to the similar process in forming the pins. Before driving the joint together or *home*, a small chamfer or corner should be taken off the points of the pins, to give more easy entrance and prevent the other piece being damaged when the pins go through, as in the mortise and tenon joint.

Half-Lap Dovetail.—This is employed chiefly for the front joints of cabinet drawers, &c., and only shows the joint to one side. The front piece should be thicker than the sides or the back. The dovetails come about two-thirds through the front piece. The pins are first drawn on the front or thick part, the

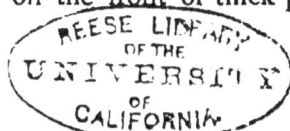

necessary distance for the depth being a little more than the thickness of the sides, and indicated by a line. The distance that the dovetails are to come through from the inside of the front piece is gauged from the inside outwards, and while the gauge is set, a line should be drawn round the end of the other piece, for the depth of the holes. The most of the work of cutting out will have to be done by the chisel, as the saw can only be used for a corner, at about the angle of 45°. When the pins are finished, the transfer must be made to the other piece, as with the plain dovetail joint, and the same care and precaution observed. The cutting out need not be again described. Figs. 67 and 68 show this joint. Glue is chiefly selected as the only fixing necessary for dovetail joints, but occasionally, when extra strength is wished, nails or screws may be added.

Wood Turning.—Lathes, used exclusively for wood turning, have no back motion, and in many instances not even cone pulleys. This is owing to the necessity for a uniformly high speed (1000 to 2000 turns per minute), and the fact that the average diameter of the work is small. In many respects there is very little difference between lathes used for wood and iron, so the description given of the parts of the latter may, to some extent, apply here. The countershaft is, however, frequently dispensed with, and in its place a slack pulley is fitted on the spindle, by the side of the driving one, so that the belt may be moved from the one to the other by hand. Wood turning lathes are provided with a variety of cup chucks, and several holding and driving adjuncts, among them the *fly* centre, which imparts the motion to the work, besides supporting it. The "T" *rest* is all that is required, as the tools are entirely worked by hand. They are of a most elementary character, the *gouge* and the *flat turning chisel* playing the most important parts. The gouge is ground on the outside, and has its corners removed, leaving a regularly curved cutting edge. Its application is varied, for by means of it the rough stock is reduced to cylindrical form, concave curves are fashioned on the surface of the work, and, by its assistance, hollow or cupped work may be performed. The flat turning chisel is ground equally on both sides to form a thin cutting edge, and at such an angle across as to make the acute corner from 45° to 60°. By means of it all flat, cylindrical, and convex surfaces are finished, and even many of the *flat* concave ones are cleaned up without difficulty. After dexterity has been acquired in the manipulation of these tools, so that flat and curved surfaces are formed with ease, many beautiful pieces of work may be produced by judicious combination. Before this stage has been reached, however, many failures will beset the path of the beginner, not the least, by any means, being the tendency of the tools to *run in* to the work. The proper position of turning chisels is obtained when the under side of the cutting edge is tangent to the surface of the wood, and when the axis of the tool is at a slightly obtuse angle to that of the lathe, the movement of the tool being towards the obtuse angle. *The chisel should be so placed that the acute corner of the cutting edge may project over the work, the only part allowed to cut being the half of the edge next the obtuse angle.* Neglect of this rule will lead to constant trouble, arising from the acute corner running into the work, which, if bad on a plain cylindrical surface, will be found much worse in turning a bead or convex surface. This is caused by the constantly varying position of the tool, and the absence of any surface on which to support it. Although easy enough to all appearance, the beginner must not be disheartened by repeated failure, for perseverance will, in the end, be rewarded by success. After the difficulties connected with the chisel have been overcome, those met with in the use of the gouge will, in most cases, disappear, because of their similarity. On no account should any of the surfaces be *scraped* with the tool at right angles, for by such means a good finish can not be obtained on *soft* materials. Very little should require to be done to turned work after the tools, unless it be a slight touch with the finest glass paper, followed by a handful of the softest turnings. Occasional oiling of the running parts, sharp tools, steady application, and a smoothly running lathe, all combine to produce good work.

PART III.

METAL WORKING.

Modern requirements have compelled the introduction of a great variety of machine tools of all sorts, whereby the tedium of production is greatly removed, and the rapidity of manufacture increased a hundredfold. By the use of turning lathes, planing, shaping, and slotting machines, together with milling and drilling machines, a never-ending variety of articles may be produced. As such are in a manner only copying machines, the exactness of the work will greatly depend on the perfection of the tool, so that although a piece of metal is shaped into cylindrical form by the use of a turning lathe, it does not follow that it will be a true cylinder, any more than that the surface produced by a planing machine will always be perfectly straight, flat, and true. Constant care must therefore be exercised in order that the slightest error may be detected at once, and, if possible, rectified.

Too much caution when engaged with any sort of power tool in motion is impossible, and unnecessarily close proximity to gearing of all kinds should be avoided. Every machine and hand tool should be carefully and thoroughly cleaned after use, and all portable parts laid away in their respective places. Constant observation of the old maxim " A place for everything, and everything in its place" should be a law in every workshop. Breakages and accidents should be reported without delay.

Turning Lathes.—The most widely employed of all machine tools is the turning lathe. By means of it, all sorts of cylindrical work are fashioned, besides a great variety of other forms. Lathes are of three general types—hand, foot, and power—depending on the source from which the motion is derived. The ordinary modern lathe is provided with a planed metal bed fixed on metal standards. Two *heads* are supported on the *bed* or *shears*, the one at the left called the *running* or *fixed head*, the other the *moving, shifting,* or *poppet head*. The running head is usually supplied with a stepped cone, corresponding to one on the counter shaft, by means of which the speed of the machine may be changed at will. The second, or *back motion,* supplied to most power lathes, is for the same purpose, and is resorted to when a very slow speed is required. The cone and spindle are fixed together by means of a screw, which must be moved before the back motion can be started. This will allow them to revolve independently. The back motion is brought into *gear* with the wheel and pinion on the spindle by the use of two eccentrics, and kept in position by a pin passing through the head. Each head is fitted with a hardened, conical, cast-steel centre, which supports, and keeps in position, shafting and such like work. The point of the running head spindle is provided with a screw, on which all *chucks* and *face plates* are fixed when in use. These lathe adjuncts are chiefly employed to hold the work in position for turning, and are of many forms. The shifting head is fixed in position by a bolt; but for small movements, the spindle may be altered by an internal screw, on which the hand wheel is fastened. When properly set, the spindle is held fast by another screw, to prevent accidental movement. Most power lathes have self-acting slide rests and tool holders combined, and are supplied with two motions—one parallel to the the axis of the lathe, and the other at right angles to it. The speed of these motions may be changed by the use of different trains of wheels, from which the movement is conveyed by a screw to the slide rest. Any lathe, having such automatic movement, is called a self-acting, screw-cutting turning lathe, for, by means of it all sort of screws may be cut with perfect regularity. All power tools ought to be provided with fast and loose pulleys, so that they may be started and stopped at will. The speed of a lathe depends on the material being turned, and its diameter; 25 feet of surface per minute being about the average. The different operations in metal turning will be most easily understood by following the production of a small cylindrical piece of iron, say 12 ins. long and 1 in. in diameter. In procuring the rough stock, sufficient allowance for turning must be made in both length and diameter. After cutting, the ends must be chipped, filed or hammered, so that the centres may be readily marked. By means of a *jenny callipers*, the centre point of each end may be found, the ends having been chalked to show the marks clearly. A small conical centre should now be formed with the *centre punch*, after which the piece of iron must be put between the lathe centres, and fixed so as to be readily revolved by one hand, while with the other, a piece of chalk should be applied to mark the high parts. This must be done at both ends, after which the piece of iron should be taken out, put into the vice end up, and the centre shifted as much as may be considered necessary. It must now be replaced between the centres, and tested again, the process being continued until the ends run true. Meantime the other parts along the length ought to be tried in the same manner, and hammered till straight. When found to run truly

through its entire length, a small hole about a $\frac{1}{16}$ in. in diameter
and an $\frac{1}{8}$ in. deep, should be drilled into the piece of iron at the
bottom of each centre, to provide for lubrication, and insure continued truth.   A driving chuck must be put upon the running
head spindle, and into it one or two pins fixed.   These pins come
into contact with the *carrier*, screwed on one end of the work, to
which the motion, imparted to the lathe, is transmitted.   After
the carrier is fixed in its place, and the work put between the
centres ready for turning, the proper tools must be selected and
sharpened.   All metal cutting tools must, like those for wood
working, have a small angle of clearance, while the cutting angles
vary from 60° to 90°, the former being suitable for wrought-
iron and steel, while cast-iron and brass require a much greater
angle.   *Roughing* tools are of an angular form at the point, and
slightly rounded ; whereas *finishing* or *planishing tools* are flat on
the face, and sharpened to an acute edge with the set-stone.
If the lathe is provided with a self-acting movement and tool
holder, then the tool must be placed, so that the cutting edge
may be about level with the centre of the work, or slightly below
it, and firmly fixed in that position.   No lathe should be set in
motion until it has been ascertained that all moving parts are clear.
This may be easily decided by turning the cone one revolution
with the hand, after which it may be started with assurance.   The
running parts should be kept well supplied with oil to prevent
heating.   Water is used in turning wrought-iron and steel to keep
the tool cold, but not in the case of cast-iron and brass.   The
turnings and waste water are caught in a box below the tool, the
water being supplied from a small can above, provided with a drip
pipe and tap.   When the tool has been brought in contact with
the work so as to give the depth of cut required, the self-acting
motion in the direction parallel to the axis of the work is started,
which will cause a cut to be removed from the surface.   This can

only be done for part of the length, as the carrier prevents the entire surface being operated upon. A series of cuts would thus be removed until within about a $\frac{1}{64}$ in. of the required diameter, which would be left for finishing. When one end is roughed down the carrier may be shifted and the other end similarly treated.

FINISHING.—The finishing or *spring* tool should now be got and fixed in the same position as the roughing tool, water being used as before. Greater care must be exercised in sharpening, setting, and fixing, than in the former process, in order to secure the best possible finish. The *callipers* ought to be occasionally applied to ascertain the diameter of the work, and so prevent mistakes, while all vibrating must be stopped if possible. For long work, this is accomplished by the use of a backstay, which is generally fixed to the slide rest and moves with the tool. The finishing cut should be light and regular, so as to produce a good surface. After the first end is done, the unfinished one will be treated in a similar manner, the catch or carrier being moved to the other end, which will have to be protected from damage by a piece of soft metal plate, such as copper, placed between the finished surface and the carrier screw. Both ends have to be turned down quite square by means of a side cutting tool, but enough material must be left about the centre to prevent the work being thrown out before it is completely finished.

POLISHING. — The highest speed the lathe possesses is employed for polishing. This process is performed by the application of emery in either of two forms, viz., as emery cloth or in a granulated condition. Both may be procured in a variety of sizes or numbers, the selection of which will depend on the quality of surface produced by the tools, and the kind of finish ultimately desired. Sometimes fine filing is introduced as an intermediate process between turning and polishing, but is not advisable. Oil has to be used along with the emery to make it cut clean and regularly. As pressure is necessary in the process, the emery is usually applied between two pieces of wood, or *legs*, hinged at one end by a strip of leather, and long enough to give the desired lever power. If the surface is not well finished at the outset, several sizes of emery would be employed, the finest being used last. By a little attention the sharp corners will be allowed to retain their acute form, which is always advisable, for the sake of appearance, unless specially desired otherwise. The necessary quality of surface having been obtained, it is then cleaned and a light coat of oil applied to prevent rusting. On no account should emery be allowed to get near any of the lathe bearings, as the effect will be very detrimental, leading to endless trouble. Should a flat surface be desired, the movement of the tool would have to be at right angles to the axis of the lathe; in other respects the details would be practically the same as in plain turning.

DRILLING IN THE LATHE.—If holes have to be bored in the lathe, the work is generally fixed to a face plate or in a chuck, and the drill or cutter bar held firmly in the tool holder. Drills, unlike turning tools, cut with two edges simultaneously on opposite sides, but the rules about their clearance and cutting angles are the same. If a large hole has to be bored from the solid, several drills of progressive sizes must be used, the last taking only a thin finishing cut, so as to produce a true hole of the required diameter. The size must be tested with the *inside callipers*, and to prevent accidents it is advisable to stop the lathe before its application. The feed of the drill would be imparted in the same manner as that of the tool in turning. Much experience and care are necessary to ensure exactness of fitting in metal, as there is very little compression possible, thus greatly differing from wood.

In the case of the foot lathe the tools have to be held by the hand, and are either straight, with short handles, or bent, and fitted with long handles. The latter are known as *heel* tools, and are used for heavy work. Hand tools are supported on a "T" rest, and the feed is produced by a combined twist and side movement of the tool handle, the heel of the tool being kept firmly in position by the left hand. Much practice is necessary to ensure good work with a foot lathe, particularly in the manipulation of the tools.

Planing Machines.—These tools are now very extensively used for producing all sorts of flat surfaces on metals. The general form is that, where the work is fixed and supported on a horizontal, moving, cast-iron table, which slides laterally on a frame having two "V" grooves, and is propelled by means of a rack and pinion. Another make of machine, much preferred by some engineers, has its table driven by a screw. The amount of *stroke* or *travel* of the plane may be changed to suit the length of the work, by moving two *horns* fixed on the edge of the table. Some machines are supplied with cone pulleys like a lathe, but many have only one speed, and may be driven direct from the main driving shaft. The power is conveyed by belt pulleys, three being needed; those on the right and left are the driving ones, whereas that in the centre is an *idle* or *slack* pulley, on which the belt must be placed when it is desired to stop the machine. The shifting of the belt is effected by a setter, moved by the two horns on the side of the table, or, if necessary, by hand. The tool is generally provided with an automatic feed-motion in three different directions, transverse, vertical, and oblique, which, combined with the motion of the table, will produce horizontal, perpendicular, and slanting surfaces. The feed-motion may be reversed by the movement of a *pawl* to right or left, and the amount of feed may be changed by altering the relative position of the feed connecting-bar ends on the bell crank levers. One peculiarity about the planing machine is its quick return motion—which is provided to save time during the idle return stroke—the cut being only taken off in one direction. This is accomplished by means of gearing.

C

The work is fixed on the table, either by bolts and plates, or by the use of a parallel jawed vice. To accommodate different sizes of work, the toolholder may be moved bodily in a vertical direction by two equally pitched screws, connected by a shaft and mitre pinions. The tools used are the same as those employed in the turning lathes, and in consequence the same rules and precautions will be necessary.

Shaping Machines.—The chief difference between these tools and planing machines is that the work remains stationary, while the toolholder has the motion given to it, and although used for similar work, still it is of a smaller, lighter, and finer character. Like the lathe, the shaping machine has a metal shears, supported on cast-iron standards, on which the toolholder slide moves. In the majority of cases the motion of the toolholder is at right angles to the shears, but recently a new machine has been introduced, in which the cutting motion is parallel to the shears. Both types of machine give satisfactory results, but the latter has unmistakable advantages for certain classes of work. Shaping machines are generally fitted with speed cones, the high speeds being used for the short cuts, and the speed decreased as the length of the stroke increases. In most machines only one self-acting motion is supplied, viz., at an angle of 90° to the direction in which the tool moves, but there are always the usual vertical and oblique movements imparted by the hand. The travel of the tool may be easily altered by moving the driving pin to or from the centre of rotation, and the position of the stroke may be changed by shifting the relation of the toolholder slide and the connecting-rod. The tools used are similar to those of the planing machine and lathe. The work is fixed to cast-iron tables (attached to the framing) by bolts and plates, &c., as in the planing machine.

Should a certain fixed amount have to be removed from a

piece of metal, either by planing or shaping machines, lines are drawn upon the surfaces at right angles to the one to be treated. As many metals are so hard on the surface that a line made by a steel marker cannot readily be seen on them, it is the usual practice in all such cases to cover the part to be marked with chalk. Even after this, the lines are often dotted along their entire length with a small sharp pointed centre punch, which will effectually preserve them. A great deal of care has often to be expended in setting and fixing particular pieces of work, the assistance of the rule, square, and spirit level being largely drawn upon.

Drilling Machines.—These tools, in their simplest forms, are used for boring holes, making recesses, and slot drilling—if supplied with a horizontal feed motion. They assume a great variety of shapes, but, for ordinary purposes, that having an upright cast-iron frame, in which a vertical spindle is fitted, capable of revolving at various speeds, and of being moved perpendicularly by means of a screw, is found most suitable. In the lower end of this spindle the drills are fixed, and at the top a hand wheel is attached by which the vertical movement is supplied. A self-acting feed-motion is used in some drilling machines, but owing to the frequency with which small drills break, it is found more advisable to apply the feed by hand. The work is supported, either by the table or bench on which the machine is fixed, on the cast-iron base of the machine itself, by a movable table connected with the machine, or on the floor. A very convenient form of machine is made where the table may be moved to various heights to suit different sizes of work. Several speeds are necessary for drilling different diameters of holes. This is obtained by the use of cone pulleys, and the horizontal motion of the driving shaft is changed into a vertical one by means of mitre pinions. Large and small holes cannot be drilled advantageously with the tool running at the same number of turns per minute,

hence arises the necessity for a variety of speeds, the greatest
velocity being used for the smallest holes, and the speed reduced
as the required diameter increases. The rule for speed, given in
metal turning, will be found suitable here also. The drill, or
cutting tool, used in this machine, is now made in a great
many different forms, all supposed to possess special important
properties. Those in most common use are the twist drill, and
the old form of flat pointed drill. Each has two cutting edges,
the angles of which must be made to suit the metals that the
holes are to be drilled in if good cutting action is desired. The
chief advantages possessed by the twist drill are the regular
discharge of the borings, the production of straight and round
holes (owing to their equal diameter throughout), and their
quicker speed of action, together with less waste of time in
sharpening. Two arguments are used against their introduction,
viz., the initial cost, and the danger of breakage through careless-
ness ; but these are more than counterbalanced by those in favour
of them. Although unequal to the twist drill, the old form of
flat drill produces a very good hole if carefully made and
sharpened, and it is still very widely employed. Both drills are
ground to a "V" shaped point, with their cutting edges on the
opposite sides. If deep holes are to be bored without the use
of the twist drill, then the machine must be stopped occasionally
so that the borings may be removed. This will prevent choking
and consequent heating, both of which are objectionable and
hurtful to the drill. Much more constant care is needed in
drilling small holes than large ones, as the slightest error will
often be followed by the destruction of the tool. In all cases
wood should be used between the table and the work, but most
particularly so if the holes have to be drilled through. The
presence of the wood will indicate when the drill reaches the
other side, and at the same time prevent the table being damaged

by the point of the tool. As the drill passes through, the feed should be reduced so as to prevent the possible breaking of the former. This often takes place, particularly if there is any slack on the spindle. The position of a hole is generally indicated by a circle, the centre of which may be found by the compass, if necessary, and fixed by means of a centre punch. Even when a beginning is made in the very middle with the drill, it is astonishing to see how far wrong it will sometimes go before it reaches the circle. When this is the case (and it will be readily detected by examination), the centre must be shifted by means of a *foals-foot* or centering chisel, having a curved edge. When a small cutting has been removed at the side most remote from the circle, the drill will have to be tried again, and the operation repeated until the result is satisfactory. By this means mistakes in drilling will be few, whereas, if carelessly treated, the trouble will be constant. A very little practice with this machine will be sufficient to learn the weight of pressure suitable for various drills and the speeds required, but, as in the case of many other tools, the sharpening of the cutting edges will prove the chief difficulty for some time.

Blacksmith Work.—The work of the blacksmith comprehends the different processes connected with the manufacture into useful and ornamental forms, of wrought-iron and steel while in a heated condition. A forge, an anvil, hammers, tongs, punches, cutting, and shaping tools are among the principal adjuncts he uses to carry on his craft. The forge, where the metal is heated, is an open fire supplied with a blast of air from a bellows, fan, or blower, to enable him to procure a high temperature readily. In close proximity to the fire, receptacles, containing coals and water respectively, should be placed. The anvil, on which the work is performed, is a suitably shaped block of wrought-iron faced with steel, and raised to a convenient height

on the end of a short log of wood let into the floor.   Forge
hammers are of three kinds—hand, sledge, and steam—to suit
different weights of work, each being of many forms.   The tongs,
punches, chisels, swedges, and shaping tools are of endless variety,
and adapted for the special branches of the trade.

Among the primary difficulties the young blacksmith meets
is the lighting of the fire, which, however, if properly gone about,
is not so difficult as it seems.   Wood shavings are very suitable
for the purpose, and in most instances can be readily procured.
When a workman leaves his forge at night, or long enough
to necessitate the relighting of the fire, he should, in all
cases, rake back the partially burnt coals, clear out all the
dross and clinkers, removing the ashes so as to expose the
mouth of the *tuyere*.   This having been properly done, a large
handful of shavings may be placed in the cavity, lit in several
places, and allowed to burn slowly until well nigh reduced to
ashes, when the partially burnt coal cinders may be cautiously
placed on them and the air blast gently applied.   More cinders
may be added as the flames catch hold, the blast may be increased,
and after a short time, fresh coals may be put on without danger.
With proper attention to these instructions failure will be next to
impossible, and there will be little trouble from the burst of smoke
unavoidable in the ordinary method.   The fire should be regularly
replenished, and should have the clinkers and other impurities
occasionally removed, so as to keep it as clear as possible.   It is
necessary to have the coals moist, to prevent undue waste, and at
the same time, to help them to adhere or *cake*, so as to form a sort
of covering or roof over the work while heating.

Iron forging is divided into three main branches—drawing-
down, upsetting, and building-up or welding.

DRAWING-DOWN is the reduction of size by the use of the
hammer on two adjacent sides alternately, in the case of square

section, the iron being twisted with the left hand, after every stroke, through a quarter of a revolution. Exactness in the amount of turning ensures the section remaining rectangular. This is very difficult for a time, and must receive constant attention until it is mastered. An approximation of flatness is troublesome to obtain for a beginner, but with watchfulness this also may be secured.

UPSETTING is the converse of the foregoing process, and is accomplished by striking on the end of the iron, so as to increase the size of section. Thè heat is applied only at the part to be upset, which may be at the end or elsewhere, and when sufficiently enlarged, the remaining processes necessary to complete the work would be performed.

BUILDING-UP OR WELDING is a much more difficult branch of the blacksmith's work than either of the foregoing, and can be done successfully only after much practice in the art. The temperature required is what is known as a *white* or *welding heat*, reached when the surfaces have begun to fuse, and it may be readily recognised by the semi-molten appearance of the iron, together with the discharge of small, vivid, white sparks. Sand must be employed as a *flux* to prevent oxidization, and to facilitate the ready removal of adhering impurities, contracted in the fire. The flux should be *sparingly* thrown by the hand on the parts to be welded while in the fire, after they have been observed to approach the proper heat. It is absolutely necessary that *both* pieces should be at a welding heat, and that no time be lost after removal from the fire, before they are brought into contact, and thoroughly welded together. A sharp stroke across the anvil with both pieces, immediately after removal from the fire, when the face of the scarf is in a vertical position, will discharge the slag from the joints, or if the pieces are too large to be so treated, it can be done by the aid of a wire brush. As the anvil is cold, the piece lying undermost will very soon lose its heat, particularly the point of the scarf. In order to

minimise the effect of this, it is a common practice to allow the
point to project for a short distance beyond the edge of the anvil,
and as soon as the upper one has been attended to, the work
would be inverted, and the other scarf welded up. For light work,
the joints used are generally of the simple oblique scarfing variety,
although several others are employed in particular cases. It is
needless to say that *bending* enters largely into blacksmith work,
as many of the beautiful effects obtained are almost entirely trace-
able to the grace of their curves. Much of this class of work is
done by the use of the anvil horn and hammer alone, although
occasionally additional means may be found advantageous.
Punching is usually performed from both sides and finished on a
steel faced *bore* of suitable size, through which the point of the
punch passes, the hole in the first instance being made smaller
than the size required, and afterwards enlarged by tapered punches.
This produces a large hole without wasting the material. The
processes performed by top and bottom swedges, and many other
shaping tools, are too numerous and varied to be even entered
upon here.

HARDENING AND TEMPERING OF STEEL.—These are very
important branches of blacksmith work, but are by no means so
difficult as usually supposed. *Hardening* simply implies heating
and rapid cooling in some cold liquid, generally water, while
tempering is the cooling of the steel at a *certain* temperature. The
latter is required so as to obtain different degrees of hardness, and
may be performed either by heating in a bath of some molten
metal which fuses at a known temperature, or by observation of
the colours that appear on steel at different degrees of heat,
ranging from 420° to 620° Fahr. The colours change from white
to yellow, straw colour, purple, and blue, each indicating a certain
degree of hardness, increasing regularly from blue to white. The
method adopted will depend on the article under treatment; some

tools, for example, have only the small edge tempered that is used for doing the work, the remainder being left quite soft, whereas in many other cases the entire piece of steel is tempered equally throughout. The tempering of a chipping chisel will explain the former process. The point of the tool is first heated to a cherry-red colour, for some two or three inches, after which it is taken from the fire and put into water, in a vertical position, for about half that distance. At this stage a very common accident, called *water-cracking* often takes place. It is caused by the sudden change of temperature at the surface of the water, which may be greatly lessened by imparting a slight vertical motion to the tool. When *quite* cold, the point should be made bright with a piece of sand-stone, in order that the colours may be seen distinctly. If everything has been done carefully and rapidly, the point should appear white, gradually changing into yellow, straw-colour, purple, and blue. These colours will now move towards the point by conduction of the heat in the body of the chisel. When the proper colour, viz., one of the shades of purple—depending on the work the chisel has to be used for—has reached the point, it should be plunged immediately into water, which will complete the operation. A little practice, so as to get acquainted with the particular kind of steel under treatment, and the proper heat required, should ensure very good work. If the steel has to be tempered uniformly throughout, then the method is quite different, the *whole* piece being hardened and thereafter *brought back* to the desired colour by borrowed heat, metal bath or 'blazing-off' with oil as commonly used for springs. Steel may be welded with care, using borax as a flux, but the greatest caution is necessary to prevent overheating, which results in the partial destruction of the steel, by liberation of the carbon. A bright red heat should in every instance be avoided, but if overheated, the damage may be partially repaired by repeated heating, and hammering till cold.

Moulding and Casting.—By means of these branches of handicraft much of our most intricate work is now performed in a comparatively simple manner. They imply the making of a mould or cavity of the desired shape in some suitable substance by means of a model or pattern, and the filling of this mould with liquid metal, which, when cold, will be of the desired form. There are three common kinds of moulding, *green sand*, *dry sand*, and *loam*. The first is the class of moulding in most general use, an explanation of which will be sufficient for our present purpose. The moulds are made in fine river sand mixed with *old foundry stock*, and in positions of close proximity to the pattern, with coal dust in the case of such metals as cast-iron, requiring a high temperature to melt them. The addition of the coal dust is to prevent the melting of the sand as far as possible, and increase the porosity of the mould. Moulds are in most instances made from patterns constructed of wood, which are formed so as to be readily extracted from the sand. An allowance of about $\frac{1}{10}$ in. per foot has to be made for the contraction of the metal in cooling, and the same amount, and often more, is given as *draw* or taper, so as to enable the patterns to be removed from the sand without damage. All patterns have, therefore, to be made so much larger than the castings required, for which cause a special contraction rule has to be worked from. Many castings are made from metal patterns, so that, in the construction of the original, double contraction must be allowed.

When the moulder has been provided with a pattern from which one or more castings may be required, and after he has been satisfied that it is properly put together, he selects suitable boxes, usually made of cast-iron and supplied with numerous cross bars, in which the sand forming the mould is held firmly together. Foundries have a large number of such boxes, generally made in pairs—the bottom one being known as the *drag,* and the top one as the *cope.* The sand is then sifted, mixed, and made moist enough for adhesion and taking the correct impression. If the pattern is small and of such a shape that it may be completely imbedded in the sand of the drag, then it should be laid face down on a flat board and the drag placed over it in an inverted position. The *facing sand,* containing the coal dust, may now be put carefully over the pattern through the bars of the drag to the depth of an inch and half or thereby and carefully rammed, more common foundry stock being added as the ramming proceeds until the box is filled flush, when the surplus should be cleaned off and the box restored to its original position. Equality in ramming is very desirable, and both extremes should be avoided. If too slack the mould will not be strong enough, but if the other extreme be the condition of the sand, the easy escape of the gases will be prevented and possible damage to the casting may ensue. A little dry sand, or brick dust, should now be shaken over the face of the drag to act as parting sand, and ensure an easy separation of cope and drag. The cope is now placed in its proper position, exact replacement being ensured by the use of guide pins, and a round pin, to form the *ingate* or entrance for the metal through the sand of the cope, is then fixed in the most suitable place. The cope may next be filled up in the same manner as the drag, and when finished the sand should be sufficiently pierced towards the mould with a sharp wire to make sure of good ventilation. In no case should the piercing point be allowed to come in contact with the pattern. *Rapping* on the outside of the boxes is necessary in all cases when part of the pattern is contained in the sand of the cope so as to slacken it, after which the top box may be removed. The upper box must be raised simultaneously at all points, and unless this is done, the mould will often suffer. The pattern may now be extracted from the sand of the drag. Occasionally the rapping outside the boxes will be all that is needed for the removal of the pattern, but in cases when the parts are deeply imbedded in the sand, a repetition of the process may be necessary when the upper box has been removed. The pattern should not be struck directly (if it can be avoided), but on a rapping-rod, often supplied for the purpose, which may also be used for lifting the pattern from the sand. The ingate pin should now be removed and an opening made into the mould. All damage done to the sand in both boxes should now be repaired, and the surfaces of the mould *sleeked* over with plumbago by the use of suitable trowels. The last-mentioned part of the work gives a better skin to the mould and makes the sand more easily removed from the casting. If small parts are found broken, the fractures may be repaired by the use of clay-water, while nails are often brought into requisition to give extra strength. The cope should now be cautiously replaced, after which the mould is ready for the metal. In many instances it will be necessary to place weights upon the top box to prevent its being raised by the molten metal. It is also a wise precaution to place something in the mouth of the ingate to prevent anything falling down before the metal has been poured. Should a casting be required hollow, clay *cores* are used to give the desired internal form. These cores are made in specially formed boxes, which, in some instances, are more difficult to construct than the patterns themselves. After removal from the core boxes, they are placed in an oven and thoroughly dried. In such castings as long water pipes, columns and similar articles, where the core might bend, small supports called *chaplets* are used. The stem is made of wrought-iron, about $\frac{1}{4}$ in. square, having on one end a piece of hoop iron of suitable shape and size. Sufficient support for the sand is not, at all times, obtained from the cross bars of the cope, particularly in large flat castings. Iron hooks, or *hangers* are very extensively employed as auxiliaries.

CASTING.—Cast-iron is melted in a furnace known as a *cupola*, but for small quantities of softer metals, crucibles of fire-clay or plumbago are made use of. For the softest metals and alloys, a ladle, heated on an open forge fire may be found quite sufficient, if large enough to contain the quantity. In any case, the utmost caution must, at all times, be taken, so as to prevent accidents, which often assume a very dangerous form. Ladles used for molten metal must be carefully prepared and made perfectly dry. This is ensured by covering the inside with clay, which has to be thoroughly dried and smoked. After everything is ready, a sufficient quantity of metal is run into the ladle, and cautiously poured into the mould, until it appears at the top of the ingate. In a short time, in the majority of cases, the metal will *set* or get into a solid state, soon after which it may be removed from the mould and exposed to the air. If there is danger of fracture, arising from unequal speed of contraction in the different parts, it is often advisable to expose only those least likely to cool quickly, and so, if possible, produce a sound casting. Sometimes the castings are left in the sand all night, so as to cool slowly. Little more remains to be done in the foundry, except to remove any surplus parts of the metal, or to drive out the cores employed in different parts of the construction.

Fitting and Finishing.—This branch of the engineer's work includes all the processes where the hammer, chisel, file, and scraper are used. These are necessarily very numerous and diversified, but not so much so as in past years, because of the introduction of so many machine tools. When small, the work is held in the vice, with soft metal grips interposed to prevent damage to the surfaces in contact with the jaws. The old form of tail vice is still very widely employed—in fact in some workshops nothing else is met with—but during recent years the constant cry about the absence of parallelism in the hold has led

to the introduction of many parallel jawed vices, some worked by
screws and others by instantaneous grip arrangements. These
are quickly growing in favour, and in many works already hold
entire sway. For medium weight of work the surfaces operated
upon should not be higher than the workman's elbow, so that he
may secure the best possible command over them. Engineers'
hammers vary chiefly in weight and form of *pane*, the latter being
made of three different kinds—cross, straight, and round. The
weight of the hammer will be determined by the strength of the
workman and the kind of work to be performed, while the size
and shape of the chisel will be guided by similar considerations.
Chisels are of several shapes, the most useful being the *flat
chipping chisel*, the *cross-cut*, and the *foals-foot* or *round nosed
chisel*. For much of the ordinary run of work, chisels are made
of $\frac{7}{8}$ in. octagonal steel, from 9 to 10 ins. long. Flat chipping
chisels, necessary for all sorts of flat work, are broad and parallel
to the point, and sharpened on the grindstone equally on both
sides, so as to make an angle of from 60° to 70° with one another
—larger and smaller angles being used when advisable. The
cross-cut is a narrow chisel broadening slightly towards the point
and sharpened like the flat chisel, though often on one side only.
It is required for cutting out flat, narrow grooves, key-ways, and
such like. The foals-foot is of the same shape as the cross-cut,
except in one respect, viz., that one of its edges is curved, and so
made suitable for cutting a semi-circular groove, and performing
such like work as centre shifting at the drilling machine. Steadiness of position and regularity of stroke in the direction of the
axis of the chisel are leading elements in easy and correct chipping.
The aim of the *fitter* is to produce as flat a surface as possible
with the chisel, and so save work with the file. This, however,
is no easy matter, for the slightest alteration of the chisel's
position will produce a corresponding variation on the surface,

the truth of which will be tested, in the first place, by steel straightedges.

FILING.—When a surface has been reduced to approximate flatness, the file is next applied.   Files vary in length, contour, and section, besides quality of cut, so that several things have to be taken into consideration in their selection.   If it is difficult to produce a flat surface on wood by the use of planes, much more so is it the case with files on metal.   Steadiness and constant watchfulness are needed to prevent any one part of a surface being reduced too low.   The ever-varying position of the file on the work increases the difficulty of maintaining the tool in a horizontal plane, particularly if the surface operated on be small.   This is caused by the overhanging ends of the file continually changing position, and, in consequence, altering the amount of lever power, which, if the pressure is not correspondingly varied, must needs give rise to the formation of a curvilinear surface.   When filing soft metals, the teeth often get filled up, and are thus prevented from working, in which case a wire brush should be freely used across the file, in the direction of the teeth.   A *true* surface plate, *slightly* smeared with red lead and oil, will have to be used as a test when the surface approaches towards completion.   The parts coming in contact with the surface plate, and therefore requiring to be filed, will be indicated by the red lead markings.   After most of the surface shows contact with the test plate, but not so regularly and equally distributed as might be desired, the file should be laid aside, and a flat steel scraper substituted.   Cutting, as this tool does, with only one point at a time, there is no difficulty in removing the exact particles desired, and by patient and careful work producing all but a perfect surface.

FINISHING.—A great deal of the finisher's work is now performed by large grindstones and buffs ; all that in many instances remains to be done being to inspect the work, and correct any irregularities, going over the surfaces with the finest emery cloth. In many surfaces absolute flatness is not required so much as perfection of finish.   In such instances the scraper would not be used at all, but files and emery of various degrees of fineness, until the surface was of the quality specified.   In order that the partial flatness of the surface obtained by the files may not be lost, the emery cloth must be applied by a piece of flat wood.   In many workshops emery is not allowed on flat surfaces intended to work on one another ; the only finish they receive coming direct from the scraper.   All finished surfaces, liable to rust, should be coated over with clean oil before being laid away ; and even this may require to be supplemented by a covering of white lead and tallow if the work be intended for transport, or to remain unused for a long period of time.

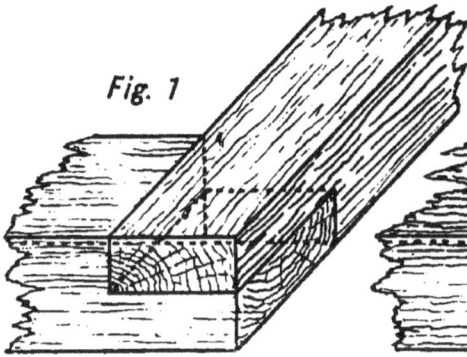

Fig. 1

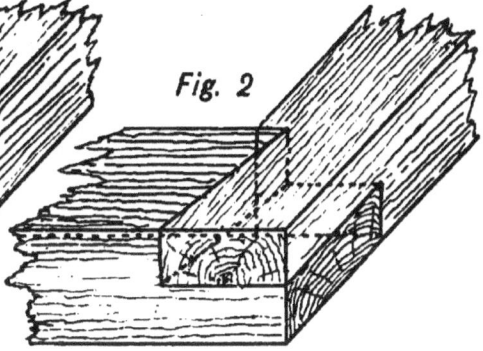

Fig. 2

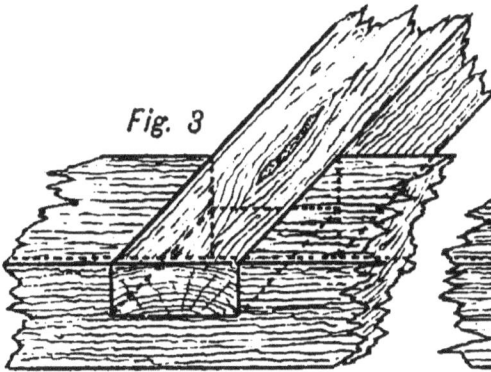

Fig. 3

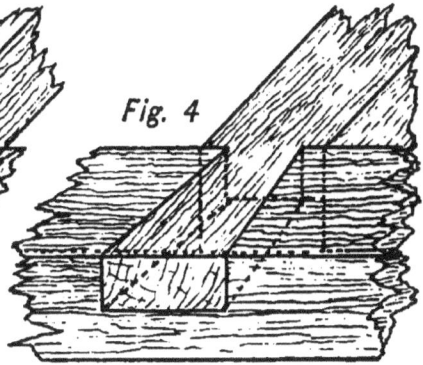

Fig. 4

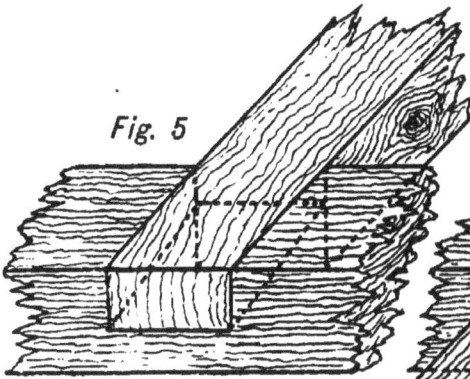

Fig. 5

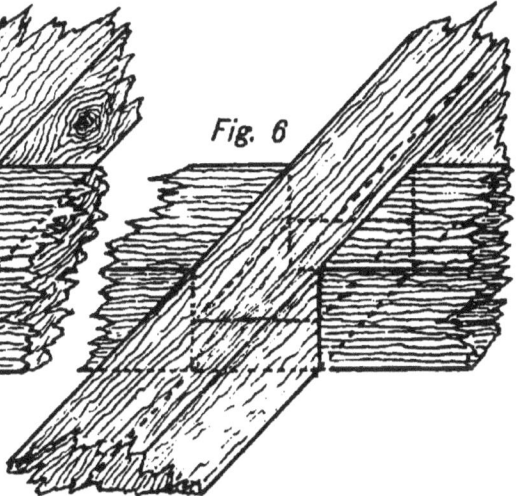

Fig. 6

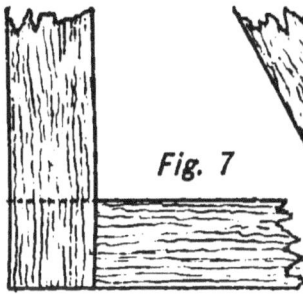
Fig. 7

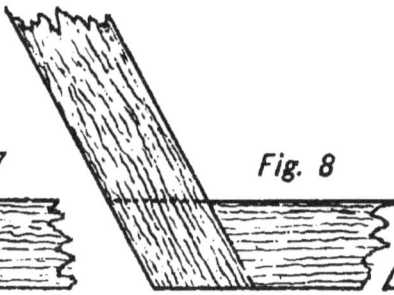
Fig. 8

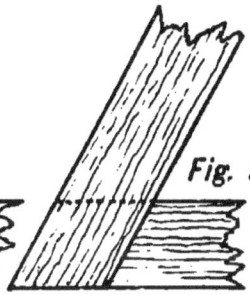
Fig. 9

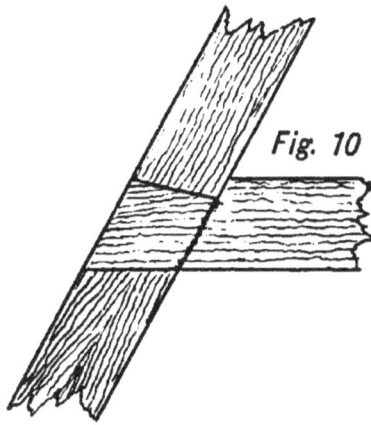
Fig. 10

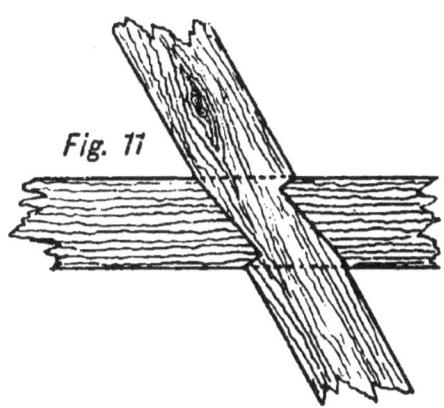
Fig. 11

Fig. 12

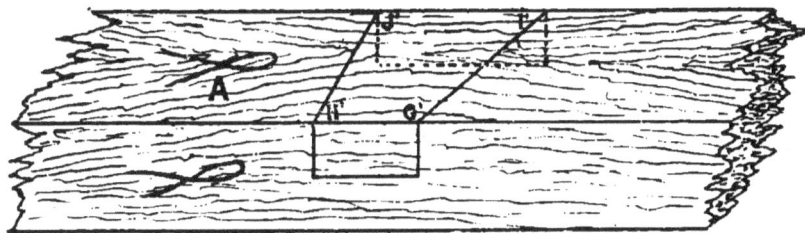

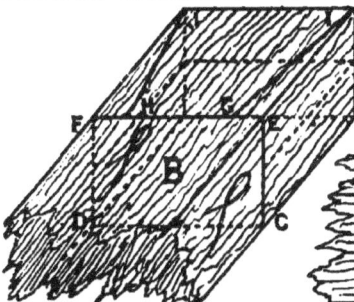

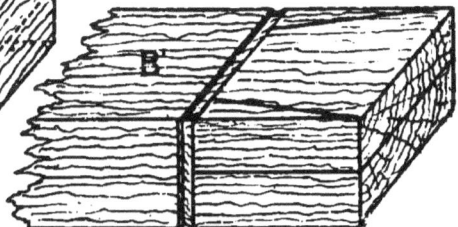

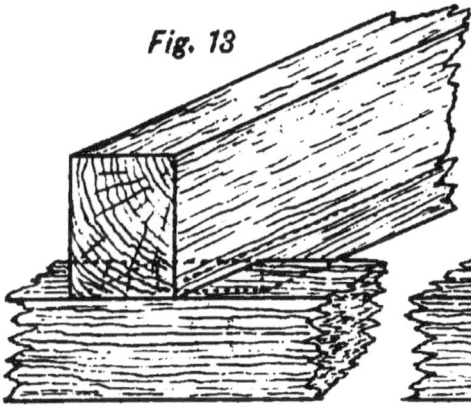

Fig. 13

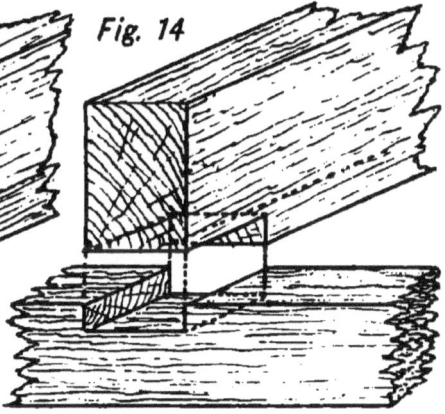

Fig. 14

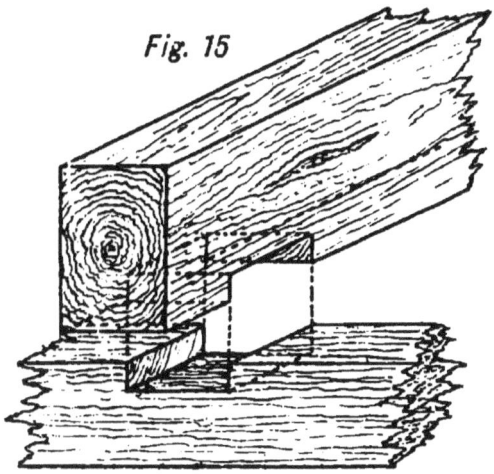

Fig. 15

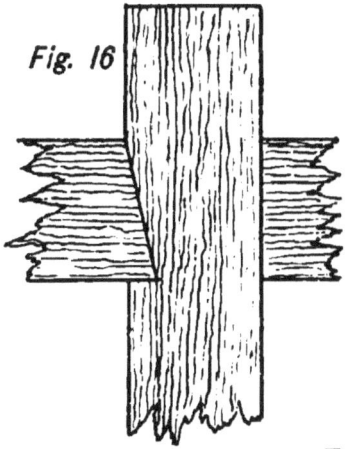

Fig. 16

Fig. 17

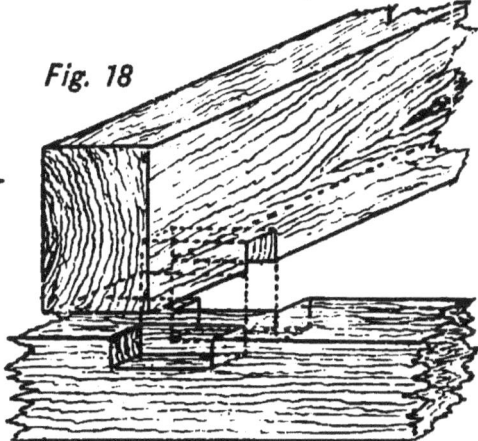

Fig. 18

Fig. 19

Fig. 20

Fig. 21

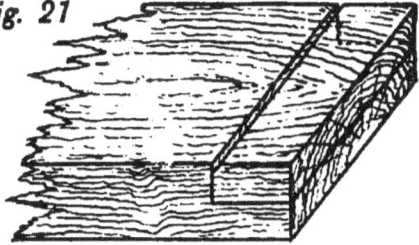

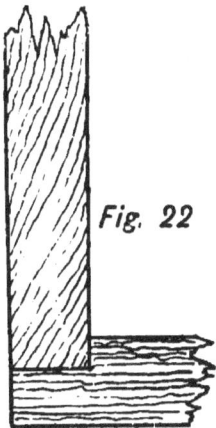

Fig. 22

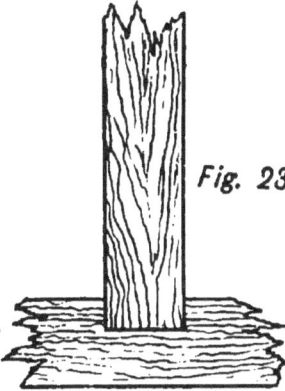

Fig. 23

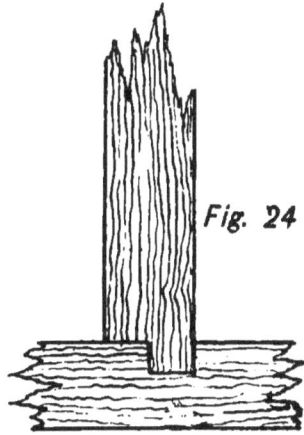

Fig. 24

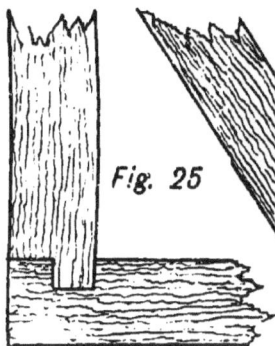
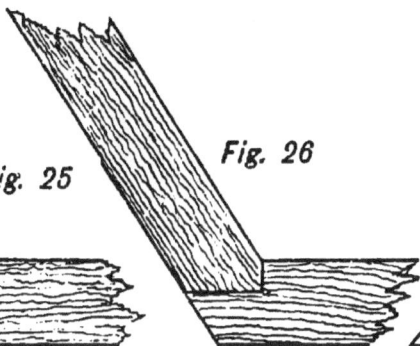
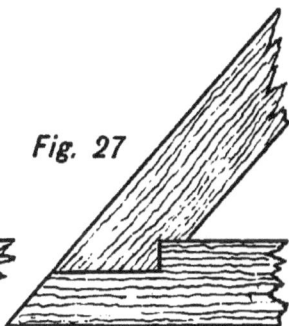

Fig. 25

Fig. 26

Fig. 27

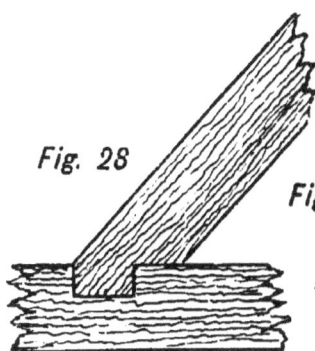
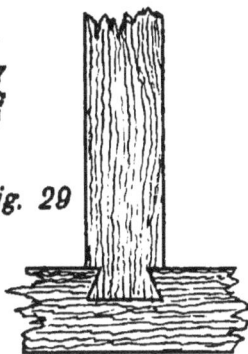
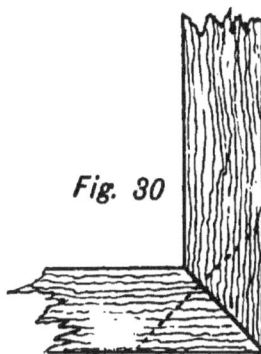

Fig. 28

Fig. 29

Fig. 30

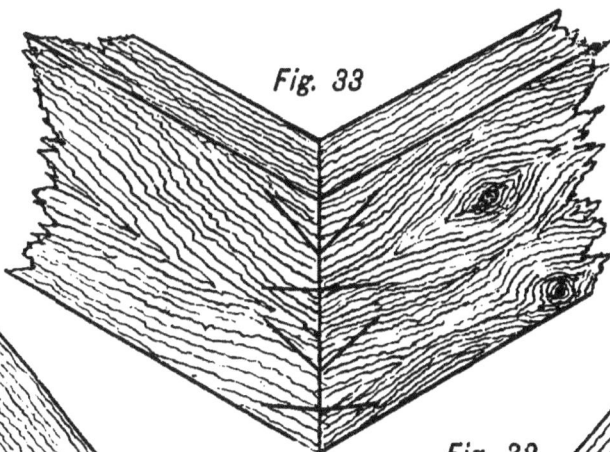

Fig. 33

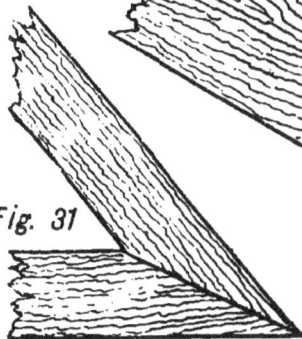
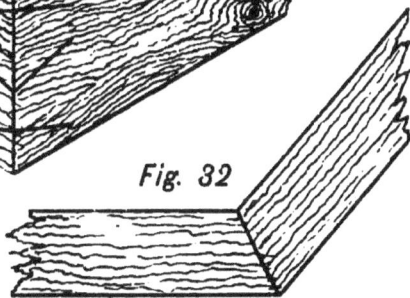

Fig. 31

Fig. 32

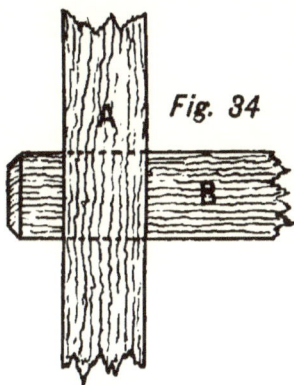

Fig. 34

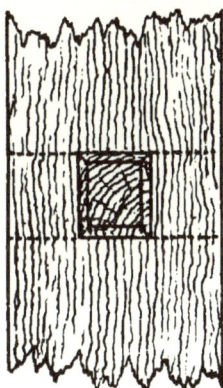

Fig. 36

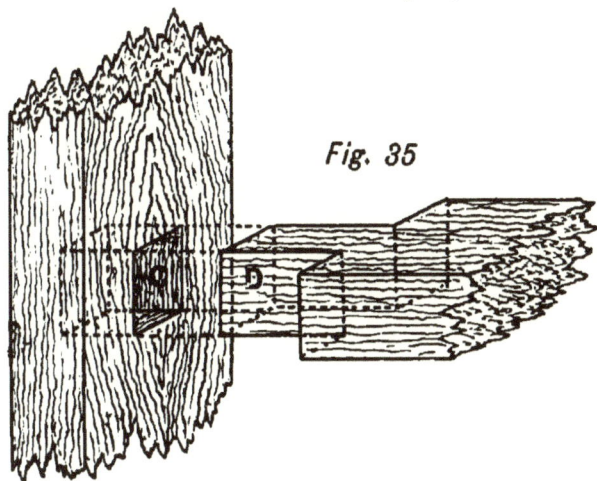

Fig. 35

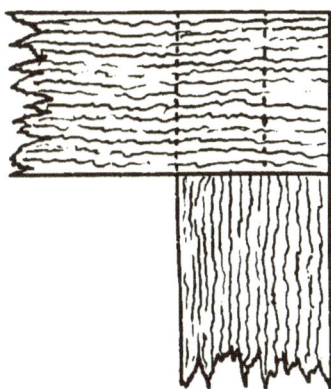

Fig. 37

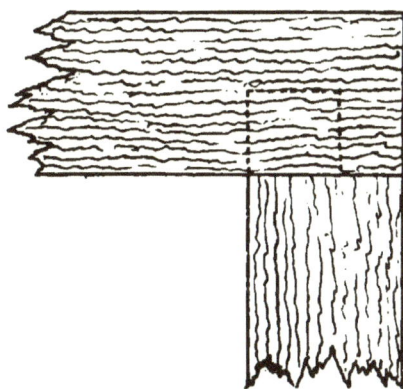

Fig. 38

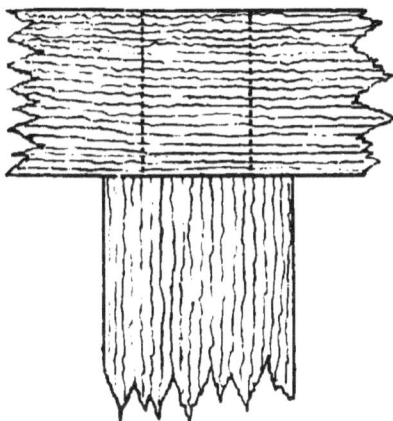

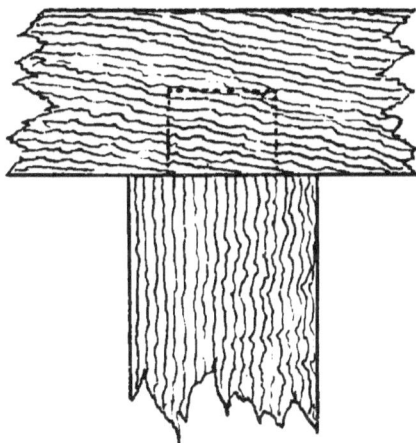

Fig. 41   Fig. 42

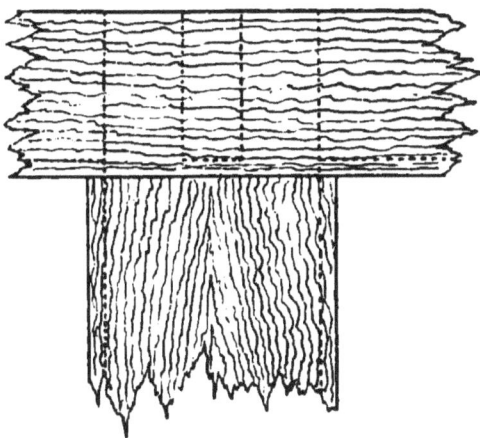

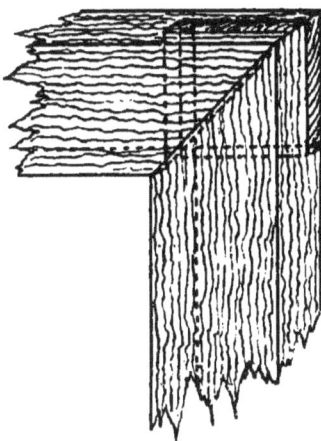

Fig. 43   Fig. 44

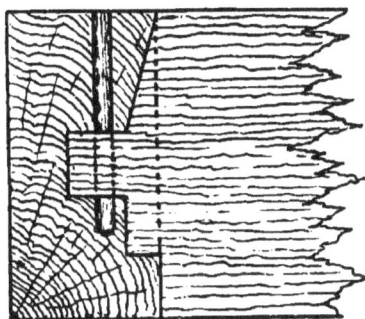

A

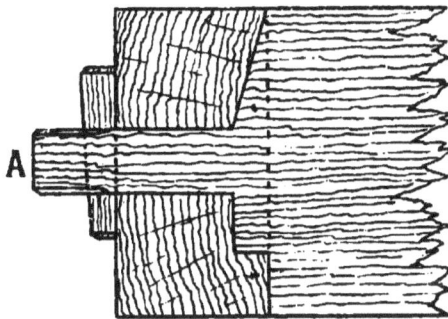

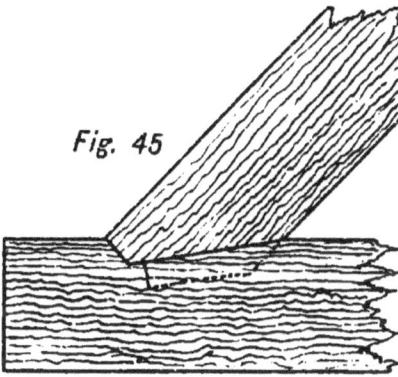

Fig. 45

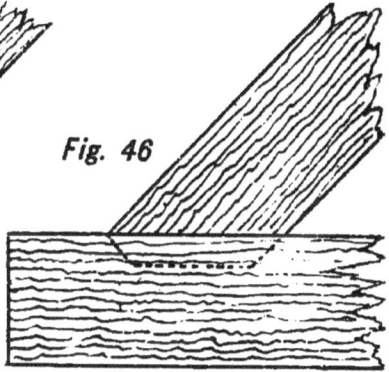

Fig. 46

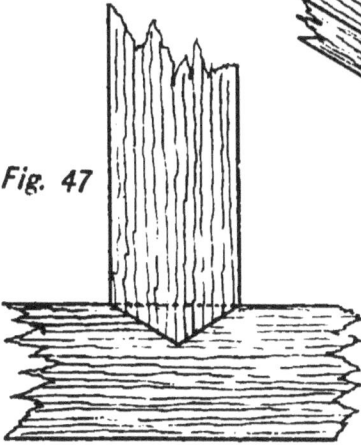

Fig. 47

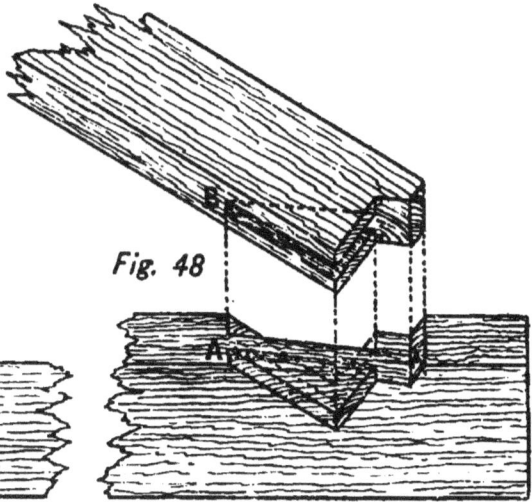

Fig. 48

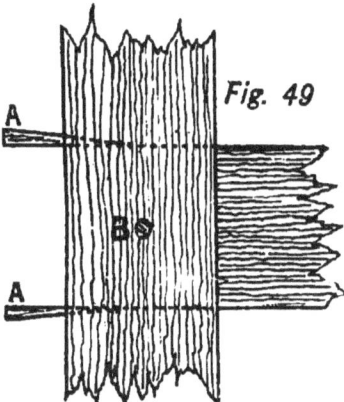

Fig. 49

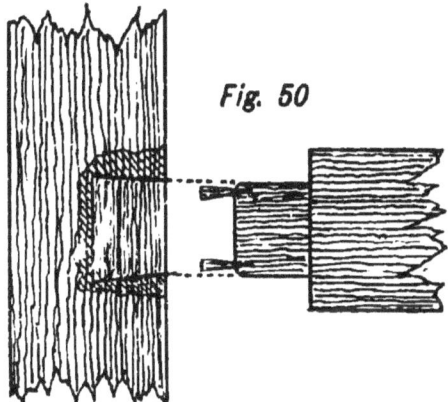

Fig. 50

Fig. 51

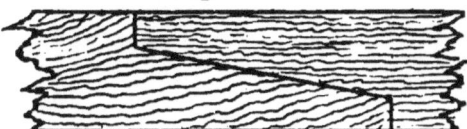

Fig. 52

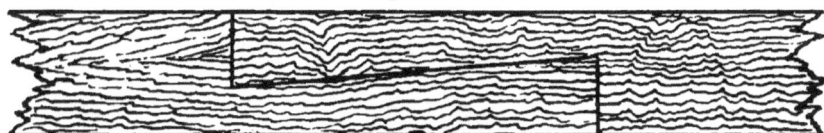

Fig. 53

Fig. 54

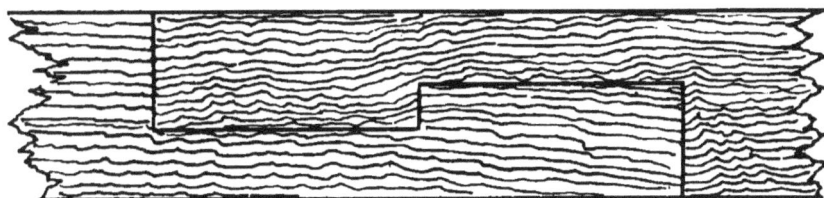

Fig. 55

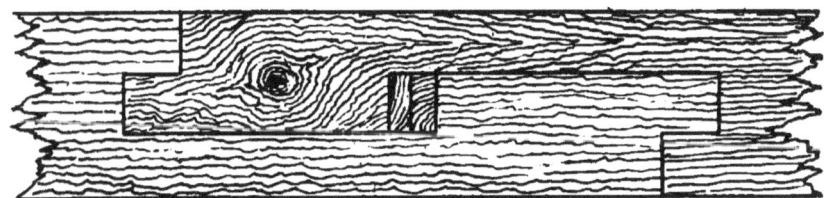

Fig. 56

Fig. 57

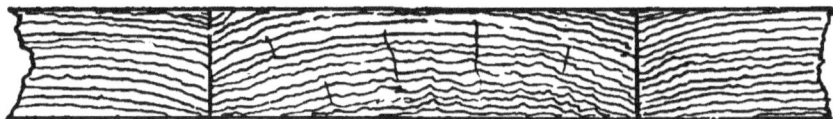

Fig. 58

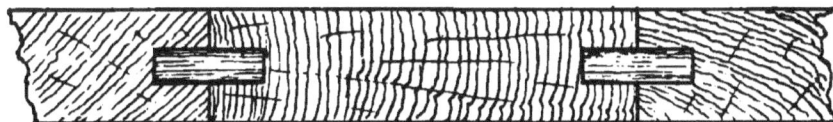

Fig. 59

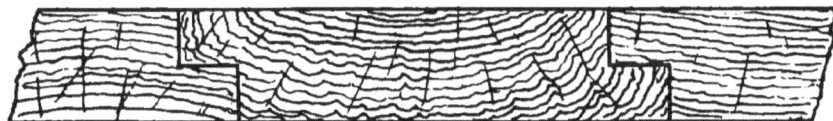

Fig. 60

Fig. 61

Fig. 62

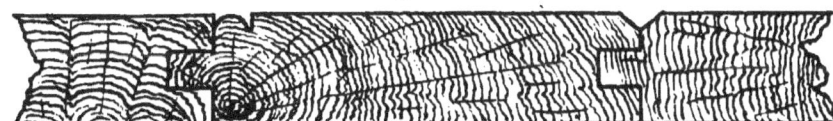

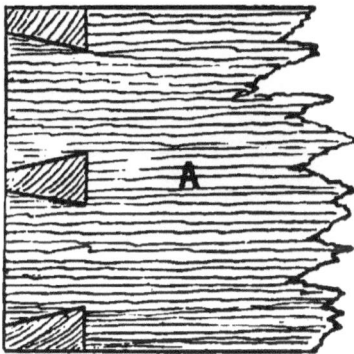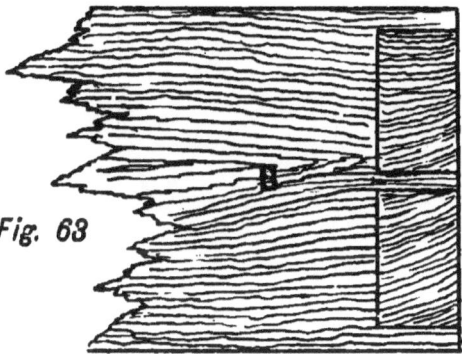

Fig. 63

Fig. 65

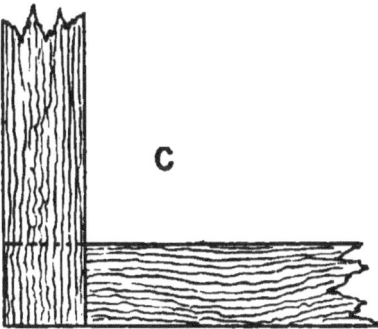

C

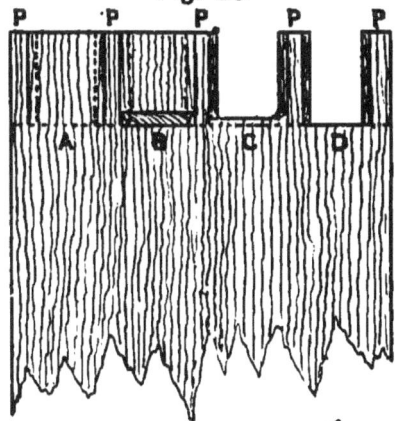

Fig. 64

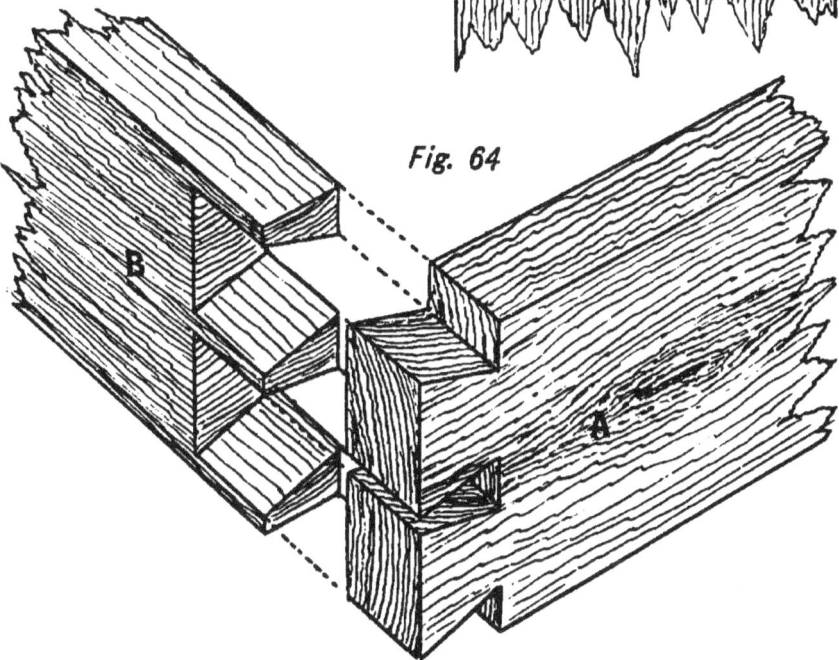

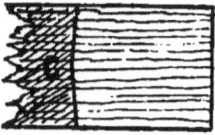

Fig. 66

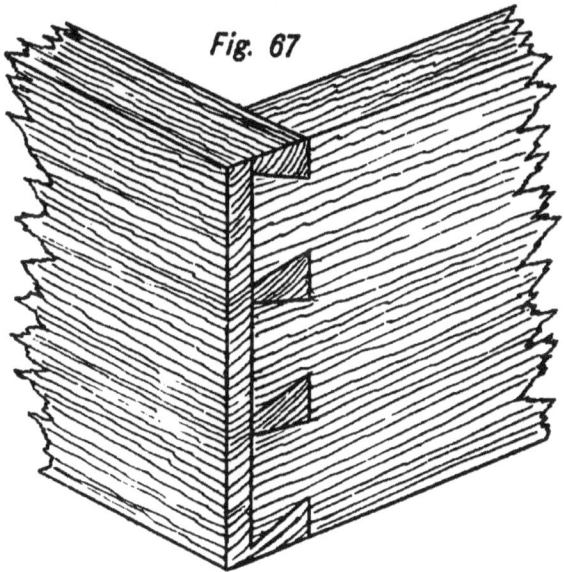

Fig. 67

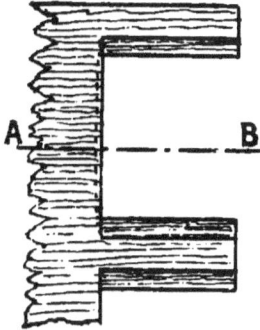

A ⎯⎯⎯⎯⎯⎯⎯ B

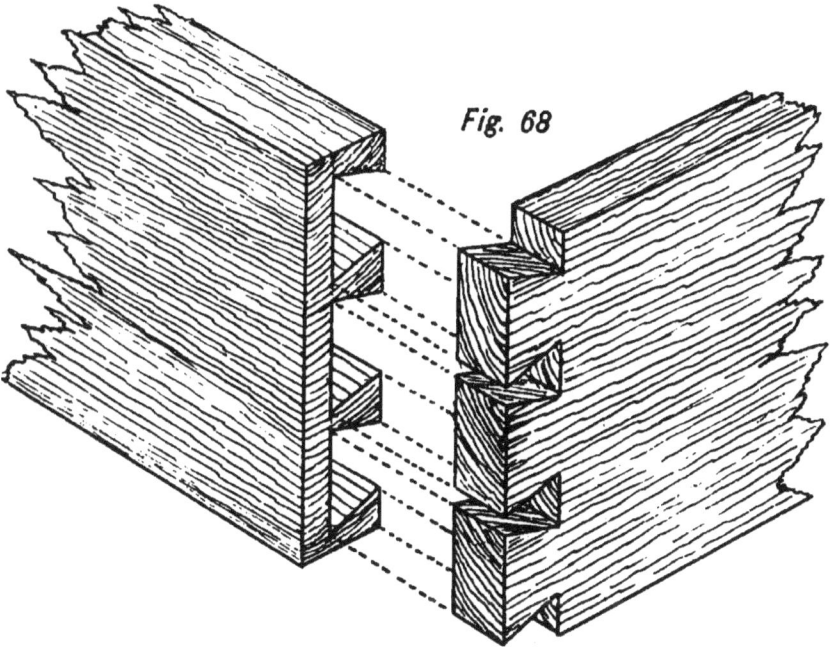

Fig. 68